Electron and Photon Impact Ionization and Related Topics 2004

Other titles in the series

The Institute of Physics Conference Series regularly features papers presented at important conferences and symposia highlighting new developments in physics and related fields. Previous publications include:

182 Light Sources 2004
Papers presented at the 10th International Symposium on the Science and Technology of Light Sources, Toulouse, France
Edited by G Zissis

181 Applied Superconductivity 2003
Papers presented at the Sixth European Conference on Applied Superconductivity, Sorrento, Italy
Edited by A Andreone, G P Pepe, R Cristiano and G Masullo

180 Microscopy of Semiconducting Materials 2003
Papers presented at the Institute of Physics Conference, Cambridge, UK
Edited by A G Cullis and P A Midgley

179 Electron Microscopy and Analysis 2003
Papers presented at the Institute of Physics Electron Microscopy and Analysis Group Conference, Oxford, UK
Edited by S McVitie and D McComb

178 Electrostatics 2003
Papers presented at the Electrostatics Conference of the Institute of Physics, Edinburgh, UK
Edited by H Morgan

177 Optical and Laser Diagnostics 2002
Papers presented at the First International Conference, London, UK
Edited by C Arcoumanis and K T V Grattan

175 Computational Accelerator Physics 2002
Papers presented at the Seventh International Conference, Michigan, USA
Edited by M Berz and K Makino

174 Compound Semiconductors 2002
Papers presented at the 29th International Symposium on Compound Semiconductors, Lausanne, Switzerland
Edited by M Ilegems, G Weimann and J Wagner

173 GROUP 24: Physical and Mathematical Aspects of Symmetries
Papers presented at the 24th International Colloquium, Paris, France
Edited by J-P Gazeau, R Kerner, J-P Antoine, S Métens and J-Y Thibon

172 Electron and Photon Impact Ionization and Related Topics 2002
Papers presented at the International Conference, Metz, France
Edited by L U Ancarini

Electron and Photon Impact Ionization and Related Topics 2004

Proceedings of the International Conference on Electron and Photon Impact Ionization and Related Topics Louvain-la-Neuve, Belgium, 1-3 July 2004

Edited by
B. Piraux

Institute of Physics Conference Series Number 183

IoP

Institute of Physics Publishing
Bristol and Philadelphia

British Library Cataloguing in Publication Data

A catalogue record for this book is available from the British Library.

ISBN 0-7503-1009-X

Library of Congress Cataloging-in-Publication Data are available

Published by Institute of Physics Publishing, wholly owned by the Institute of Physics, London
Institute of Physics Publishing, Dirac House, Temple Back, Bristol BS1 6BE, UK
US Office: Institute of Physics Publishing, The Public Ledger Building, Suite 929, 150 South Independence Mall West, Philadelphia, PA 19106, USA

Printed in the UK by Short Run Press Ltd, Exeter

Preface

This book is a collection of the contributions of most of the invited speakers to the International Conference on "Electron and Photon Impact Ionization and Related Topics" which was held in Louvain-la-Neuve, Belgium, from 1st to 3rd July 2004.

This meeting which takes place every year brought together about 80 scientists, from 19 different countries, working in the field of simple and multiple ionisation by electron and photon impact as well as related topics. The program included 22 invited talks and 54 posters on recent experimental, theoretical and computational achievements in the realization, interpretation and modelling of correlated processes involving a wide range of targets : atoms, molecules, clusters or surfaces. Beside the regular talks, three review talks have been organized on three hot topics: anisotropy and polarization in Auger-electron emission, multiple ionisation of atoms in strong fields and theoretical and practical aspects of photoionization with excitation.

Financial support from the "Faculté des Sciences de l'Université catholique de Louvain" as well as from the "Fonds National de la Recherche Scientifique de la Communauté Française de Belgique" is gratefully acknowledged. This financial support permitted the participation of young scientists from eastern and Latin American countries.

I thank the members of the International Scientific Committee and my colleagues, in particular Xavier Urbain, as well as the students from the "Laboratoire de Physique atomique, moléculaire et optique de l'Université catholique de Louvain" for their help in the organization of this meeting.

I am also indebted to Ugo Ancarani from the Université de Metz who shared with me his experience in organizing such meeting.

Finally, I want to thank all speakers who contributed to these proceedings and all participants who gave their valuable time in carefully refereeing all contributions.

B. Piraux
November 2004

International Scientific Committee

L Avaldi (Italy)

H Bachau (France)

K Bartschat (USA)

J Berakdar (Germany)

A Crowe (United Kingdom)

C Dal Cappello (France)

R Doerner (Germany)

A Huetz (France)

M Kwato Ndjock (Cameroon & Italy)

A Lahmam-Bennani (France)

P Lambropoulos (Greece)

B Lohmann (Australia)

D Madison (USA)

L Malegat (France)

Y Popov (Russia)

R Rivarola (Argentina)

A Starace (USA)

G Stefani (Italy)

J Ullrich (Germany)

E Weigold (Australia)

Contents

Inst. Phys. Conf. Ser. No. 183
Paper presented at Int. Conf. Electron and Photon Impact Ionization and Related Topics, Louvain-la-Neuve, 2004

Identification of double ionization mechanisms, results bearing on electron correlation measurements

M A Coplan
Institute for Physical Science and Technology, University of Maryland, College Park MD
R W van Boeyen
Deprtment of Chemistry, Johns Hopkins University, Baltimore MD
J H Moore
Department of Chemistry and Biochemistry, University of Maryland, College Park MD
J P Doering
Department of Chemistry, Johns Hopkins University, Baltimore MD
J W Cooper
Institute for Physical Science and Technology, University of Maryland, College Park MD
N Watanabe
Institute for Molecular Science, Okazaki 444-8585, Japan

Abstract

Electron impact double ionization of atoms and molecules offers a way to directly measure two-electron correlation, provided a number of conditions are satisfied. The ionizing collisions should take place in the binary, impulsive regime so that all of the energy and momentum lost by the projectile electron in each of the collisions is fully transferred to the target electrons. It is also necessary that the target electrons acquire sufficient kinetic energy from the ionizing collisions to leave the target atom or molecule in times short compared to an electron orbital period. Finally, the double ionization mechanisms must be well understood and separated from interfering processes. Measurements of the double ionization cross-sections for the two 3s electrons of magnesium have led to the identification and separation of two double ionization mechanisms. From the measurements the means by which two-electron momentum distributions can be obtained are clear.

1. Introduction

The most direct experimental insight into electron correlation is through measurement of two-electron probability densities, the square modulus of the two-electron wave function. Ezra and Berry (1983) and Krause and Berry (1985) first clearly demonstrated the sensitivity of two-electron position probability densities to electron correlation in atoms and were able to assign, in a qualitative way, correlated modes of motion to different states of two-electron atoms. Krause et al. (1987) demonstrated the equivalent sensitivity of two-electron momentum probability densities. These results established the foundation for measurements of electron correlation and a rigorous test of theories and calculations. As an example we show in Fig. 1 the results of calculations of two-electron momentum densities for the pair of 3s electrons in magnesium. The figure shows plots of the probability of finding one of the electrons with a momentum $\mathbf{q}_1$ of a fixed magnitude 0.0, 0.5, and 1.0 a.u. and the other electron with a momentum (q_x, q_y) where q_x is the component of momentum parallel to $\mathbf{q}_1$and q_y is the component of momentum perpendicular to $\mathbf{q}_1$. The calculations are based on a simple 5-term CI wave function.

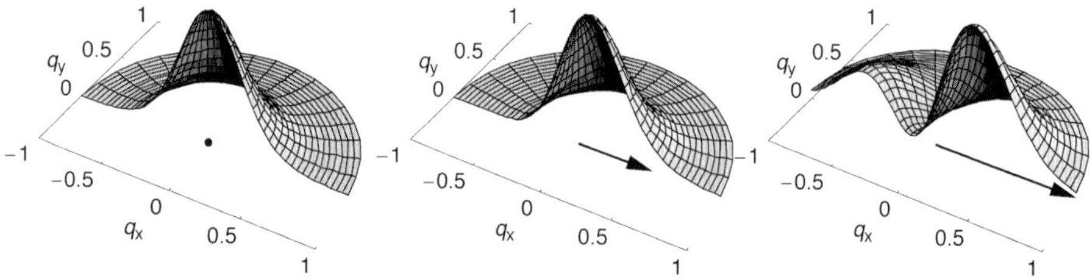

Figure 1. Momentum densities for the 3s electrons of magnesium. The momentum vectors for the electron with fixed momentum of 0.0, 0.5, and 1.0 a.u. are shown pointing in the positive q_x direction.

These model calculations show that two-electron momentum densities are sensitive to the correlated motion of electrons. The determination of two-electron momentum densities requires that the momentum of one of the electrons be determined for a fixed magnitude and direction of the momentum of the second electron. These measurements must be done in a way that does not perturb the motion of the electrons in the atom; they must be the equivalent of "snap shots" of the atomic electrons.

2. Double ionization

Electron-impact ionization as a nonperturbative probe of electron motion was first demonstrated in the 1970's (McCarthy and Weigold 1976). Extensive experimentation demonstrated that the cross section for electron-impact ionization with relatively high-energy (1 keV) projectile electrons, large momentum-transfers (2-5 a.u.) and impulsve collisions can be accurately separated into the product of a scattering term and a target structure term (Pinkás et al. 1998). Single electron momentum densities have been extracted from measurements of the (e,2e) cross sections for a wide variety of atoms and molecules (Coplan et al. 1994, 1996).

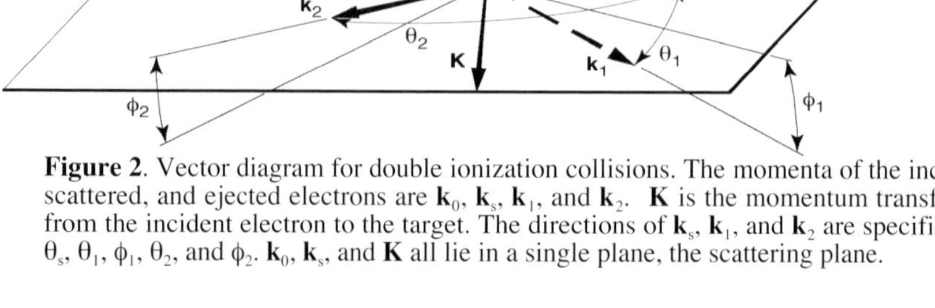

Figure 2. Vector diagram for double ionization collisions. The momenta of the incident scattered, and ejected electrons are $\mathbf{k}_0$, $\mathbf{k}_s$, $\mathbf{k}_1$, and $\mathbf{k}_2$. $\mathbf{K}$ is the momentum transferred from the incident electron to the target. The directions of $\mathbf{k}_s$, $\mathbf{k}_1$, and $\mathbf{k}_2$ are specified by θ_s, θ_1, ϕ_1, θ_2, and ϕ_2. $\mathbf{k}_0$, $\mathbf{k}_s$, and $\mathbf{K}$ all lie in a single plane, the scattering plane.

Anticipating the success of (e,3e) experiments, Glassgold and Ialongo (1968), Smirnov et al. (1978), Amusia (1993), Popov et al. (1994), and others speculated that if two electrons were simultaneously knocked out of a target atom, the two-electron momentum density could be measured and used to learn about the correlated motions of the electrons in the atoms. The double ionization (e,3e) process is described in the vector diagram shown in Fig. 2.

If two atomic electrons are ejected in an impulsive fashion, it is, in principle, possible to determine information about their initial momenta, $\mathbf{q}_1$ and $\mathbf{q}_2$, from the measurement of their final momenta, $\mathbf{k}_1$ and $\mathbf{k}_2$, and a knowledge of the scattering kinematics ($\mathbf{k}_0$, $\mathbf{k}_s$, and $\mathbf{K} = \mathbf{k}_0 - \mathbf{k}_s$, the incident-electron momentum, the scattered-electron momentum and the transferred momentum). (e,3e) experiments for the investigation of correlated electron motion pose difficulties and complications in addition to those encountered in (e,2e) experiments. To start with, it is necessary to know the mechanism by which the two electrons are ejected from the target in order to calculate their initial motion from measurements of their motion following ejection. Beyond this fundamental requirement of many-body mechanics, there are practical experimental difficulties. To achieve the necessary conditions it is necessary that the energies of the ejected electrons exceed their binding energies by a factor of 2 to 3 at the minimum, however under impulsive conditions the (e,3e) cross section decreases as the inverse third power of the ejected-electron energies, so the signal from the experiment will decrease rapidly as the requisite ejected electron energies and momentum transfer are approached.

There have been a number of (e,3e) measurements (Dorn et al. 2002, 2003, Taouil et al. 1998, Lahmam-Bennani et al. 1989, 1991), but until recently, no fully impulsive (e,3e) measurement has been reported, nor has it been possible to unambiguously identify and isolate an impulsive mechanism that would be suitable for the measurement of two-electron momentum densities. The situation has changed recently, as results have become available from an (e,3e) apparatus that has been operating in our laboratory (van Boeyen et al. 2004). A particular feature of this device is that (e,3e) cross sections can be measured both in and out of the scattering plane. Moreover, this apparatus is multiplexed in such a way that up to 28 triple-coincidence measurements can be carried out and recorded simultaneously. Within the past two years the sensitivity of the apparatus has been increased by over two orders of magnitude while the background has been reduced by a factor of ten. We have reported the cross sections measurements for the impulsive ejection of two 62 eV electrons from magnesium by 1059 eV electron impact (van Boeyen et al. 2004); here we report additional measurements for the ejection of two 120 eV electrons by 2175 eV incident electrons. These results establish, we believe, the practical feasibility of the (e,3e) technique for the study of the motion of pairs of correlated electrons in atoms.

3. Impulsive (e,3e) with identification of mechanisms

The application of (e,3e) electron-impact double-ionization to the determination of the two-electron momentum densities requires experimental conditions that restrict the mechanism to two successive single-ionizing collisions of the incident electron with each target electron. Each of the single-ionizations must satisfy impulsive collision conditions with high incident–electron energy and large momentum-transfer. When these conditions are achieved, and the incident-, scattered-, and ejected-electron momenta are determined for each collision, the perpendicular component of the momentum of each ejected electron at the instant of its ejection from the target can then be obtained by subtracting the momentum transferred to the ejected electron from the measured ejected-electron momentum. Our experiment has been designed to isolate mechanisms of double ionization suitable for use in the experimental measurement of two-electron momentum densities.

The two pathways for impulsive, doubly-ionizing collisions are TS1, where the incident electron interacts with only one of the target electrons that in turn interacts with the second (Tweed, 1992), and TS2, where the incident electron interacts with each electron in sequence. While effects observed in fully differential cross sections have been attributed to TS1 and TS2 (Dorn et al. 2002, 2003, Jia et al. 2002, 2003, El-Marji et al. 1997, Lahmam-Bennani et al. 2003), we have sought to measure TS1 and TS2 ionization involving impulsive, binary collisions. In the analysis below we develop the basis for our experimental measurements of TS1 and TS2. For simplicity we neglect the binding energies of the target electrons, assume that they are stationary at the instant of collision, and that they have equal ejected energies.

Following the first collision in the TS1 mechanism, the incident electron is scattered with a momentum $\mathbf{k}_s$, transferring momentum $\mathbf{K}$ to the first target electron. The $\mathbf{K}$ vector is perpendicular to $\mathbf{k}_s$. The first ejected electron, now with momentum $\mathbf{K}$, strikes the second target electron and, sharing its momentum, is scattered with momentum $\mathbf{k}_1$. The second target electron is ejected with momentum $\mathbf{k}_2 = \mathbf{K} - \mathbf{k}_1$. For equal ejected-electron energies, both $\mathbf{k}_1$ and $\mathbf{k}_2$ will lie in a plane that contains $\mathbf{K}$ with the angles between $\mathbf{k}_1$ and $\mathbf{K}$ and $\mathbf{k}_2$ and $\mathbf{K}$ both equal to 45°. This plane can assume any orientation about $\mathbf{K}$.

After the first collision in the TS2 mechanism, the incident electron is scattered with a momentum $\mathbf{k}_s'$, transferring momentum $\mathbf{k}_1$ to the first target electron. In the second collision, the incident electron, now with momentum $\mathbf{k}_s'$, is scattered with momentum $\mathbf{k}_s$ transferring momentum $\mathbf{k}_2$ to the second target electron. The three scattering planes defined by $\mathbf{k}_0/\mathbf{k}_s'$, $\mathbf{k}_s'/\mathbf{k}_s$, and $\mathbf{k}_0/\mathbf{k}_s$ are constrained to form an irregular tetrahedron with the fourth side bounded by $\mathbf{K}$, $\mathbf{k}_1$ and $\mathbf{k}_2$. Under these circumstances, the $\mathbf{K}/\mathbf{k}_1/\mathbf{k}_2$ plane can have only a fixed orientation with respect to the $\mathbf{k}_0/\mathbf{k}_s$ scattering plane. This means that TS2 events can only be observed by detectors that lie in the $\mathbf{K}/\mathbf{k}_1/\mathbf{k}_2$ plane.

The geometry and kinematics of the experiment were chosen to accommodate electrons ejected from the target atom by both the TS1 and TS2 mechanisms. The target atom was magnesium, a pseudo-two-electron atom with two 3s electrons outside of a closed shell. Identical ejected electron energies of 62 and 120 eV were chosen to meet the requirement of being significantly greater than their binding energies. These energies also fall within the energy regions where Auger electrons are absent. Table I lists the kinematics that were used with a scattering angle of 20° and a momentum transfer direction of –70° with respect to the incident electron direction.

Table I

E_0 (eV)	E_s (eV)	E_1=E_2 (eV)	k_0 (a.u.)	k_s (a.u.)	k_1=k_2 (a.u.)	K (a.u.)
1059	912	62	8.8	8.2	2.1	3.0
2175	1912	120	12.6	11.8	3.0	4.3

An incident electron current of 1 to 2 μA was used. Scattered electrons were collected by a tandem hemispherical analyzer with an entrance aperture ±2.25° by ±2°, and an energy resolution of 12 eV FWHM. Ejected electrons with momentum directed 45° from $\mathbf{K}$ were collected by a single truncated spherical analyzer with symmetry axis coincident with $\mathbf{K}$ and detected by an array of eight detectors as illustrated in Fig. 3. Detector locations are labeled by the angular position about **K** relative to the ($\mathbf{k}_0$, $\mathbf{k}_s$) scattering plane. The energy resolution was 8 eV FWHM and the analyzer entrance apertures accepted electrons over ±7° by ±4.5° about the nominal scattering angle. There are 28 possible pair combinations of ejected-electron detectors, all of which are monitored simultaneously in triple coincidence with the scattered electron with nanosecond timing resolution. The maximum observed triple-coincidence rate for E_0 = 1059 eV was 4 × 10^{-4} events/s. In this geometry for the energies and momenta listed in Table I, the signature of TS1 ionization is triple coincidence events between ejected electron detectors separated by 180°. TS2 ionization gives triple coincidence events only for the detectors at 90° and 270°.

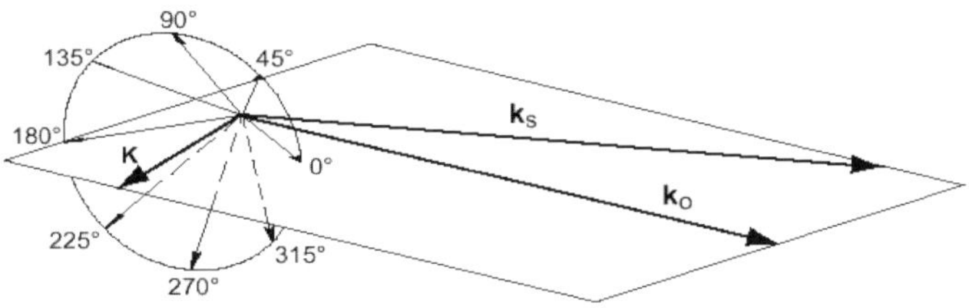

Figure 3. Experimental geometry for the observation of impulsive high momentum transfer double ionization collisions.

The measured fully differential differential cross sections for the $3s^2$ double ionization of magnesium at E_0 = 1059 eV and E_0 = 2175 eV for the geometry of Fig. 3 are shown in Fig. 4. The heights of the vertical panels are proportional to the magnitudes of the cross sections and the panel orientations indicate the two ejected-electron detector angles for which the cross sections were measured. The cross sections were placed on an absolute scale with an accuracy of ±50% using known single ionization cross sections.

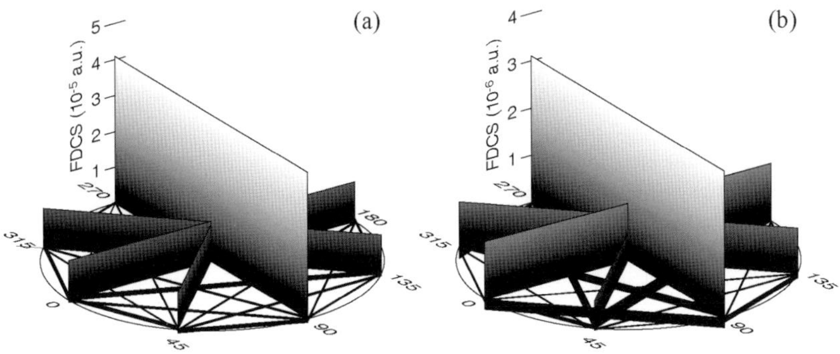

Figure 4. Double ionization five-fold differential cross sections (FDCS) as a function of ejected electron angle pairs. a. 1059 eV incident electron energy and 62 eV ejected electron energy. b. 2175 eV incident electron energy and 120 eV ejected electron energy.

For both incident energies the (e,3e) cross sections are very small for detector pairs not separated by 180°. This means that for the geometry and kinematics of this experiment, there are no ionization mechanisms that do not involve discrete impulsive collisions. In particular, Auger and shake-off processes that would give signals between all pairs of ejected electron detectors do not contributing to double-ionization in this regime. The cross sections for the (0°, 180°), (45°, 225°), and (135°, 315°) pairs are all of the same magnitude within the experimental uncertainties and correspond to the ejection of a pair of electrons with momentum vectors coplanar with the momentum-transfer vector. The mechanism for the ejection of these electrons is identified as TS1.

The cross sections are largest for the pairs of electrons ejected in the planes that are perpendicular to the scattering planes, that is, one electron detected at 90° and the other at 270°. The (90°, 270°) cross section appears to result from a combination of TS1 and TS2 ionization. Assuming a quarter of the 90°/270° pairs of electrons are ejected by the TS1 mechanism, the remaining three quarters can be identified with TS2. For E_0 = 1059 eV, the TS1 cross section has a value of 1.1×10^{-5} a.u.; for E_0 = 2175 eV the average value of this cross section is an order of magnitude smaller. Subtracting the TS1 values from the (90°, 270°) values gives 3×10^{-5} a.u. for the TS2 cross-section at E_0 = 1059 eV and 2×10^{-6} a.u. at E_0 = 2175 eV. For this experiment, assuming impulsive, binary collisions, the cross sections for both the TS1 and TS2 ionizations should scale approximately as the energy of the ejected electrons to the inverse third power. This gives a ratio of the TS1 and TS2 cross sections at E_0 = 1059 eV to those at E_0 = 2175 eV of 0.12 compared with the experiment values of 0.076. The calculated ratio of the TS1 to TS2 cross sections for the kinematics of the experiment is 0.25. From the experimental measurements we infer a value of 0.24 We believe that the measured angular dependence of the (e,3e) cross sections along with the measured energy dependence and relative cross sections provide strong evidence for the TS1 and TS2 mechanisms and the binary, impulsive collision regime.

4. TS2 and two-electron momentum densities

The TS2 mechanism is the appropriate process for the investigation of the relative motion of pairs of atomic electrons because information about the momenta of the ejected electrons in the atom is preserved in the course of the two sequential collisions. In this regard it is important to realize that the angle between the ejected-electron momentum vectors is approximately a right angle for the TS1 mechanism, while for TS2, the angle between the ejected-electron momentum vectors can be varied from about 0° to 180° simply by changing the choice of the incident-electron scattering angle. It is thus possible to isolate ionization by TS2 from all other ionization processes.

To illustrate the way in which the TS2 mechanism can be used to measure two-electron momentum densities we let $\mathbf{K}_1$ and $\mathbf{K}_2$ be the momenta imparted to the two target electrons in each collision with the incident electron. If the two target electrons are stationary at the instant of ejection from the atoms, the electrons will be ejected with momenta $\mathbf{k}_1 = \mathbf{K}_1$ and $\mathbf{k}_2 = \mathbf{K}_2$. If, on the other hand, electron 1 initially has momentum $\mathbf{q}_1$ and electron 2 initially has momentum $\mathbf{q}_2$, the electrons will be ejected with momenta $\mathbf{k}_1 = \mathbf{K}_1 + \mathbf{q}_1$ and $\mathbf{k}_2 = \mathbf{K}_2 + \mathbf{q}_2$. For ejected-electron energies selected such that $|\mathbf{k}_1| = |\mathbf{K}_1| = |\mathbf{k}_2| = |\mathbf{K}_2|$, as shown in Fig. 5, the initial momenta of the electrons can be determined from the relations $|\mathbf{q}_1| = 2|\mathbf{K}_1| \sin(\beta_1/2)$ and $|\mathbf{q}_2| = 2|\mathbf{K}_2| \sin(\beta_2/2)$, where β_1 and β_2 are the detection angles of the ejected electrons relative to $\mathbf{K}_1$ and $\mathbf{K}_2$. From the geometry of Fig. 5, it can be seen that the measured momenta, $\mathbf{q}_1$ and $\mathbf{q}_2$ can be either parallel or anti-parallel. This is the most useful case to study as it corresponds to conditions where correlation effects are expected to be greatest and most clearly reflect the nature of the correlated electron motion (see Fig. 1). Two-electron momentum densities can be obtained by recording triple-coincidence rates for detection of ejected electrons at a range of angular positions about $\mathbf{K}_1$ with ejected electrons at a range of angular positions about $\mathbf{K}_2$. The geometry is similar to that of Fig. 3, however, to exclude non-TS2 ionization events, the included angle between $\mathbf{k}_1$ and $\mathbf{k}_2$ is

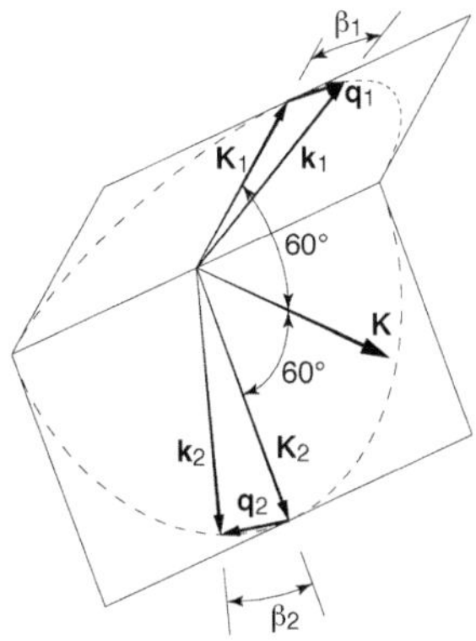

Figure 5. Double ionization kinematics for two-electron momentum density measurements.

increased from 90° to 120°. The signal from the desired sequential double-ionization mechanism will not be affected by this change in included angle. There is a direct relationship between the measured (e,3e) cross section and the two-electron momentum density. This is shown in the model calculations of Fig. 6 in which the (e,3e) cross section is shown as a function of detection angle β_2 for a fixed value of detection angle β_1 for the geometry of Fig. 5. The conditions are E_0 = 1000 eV, E_s = 912 eV, E_e = 62 eV, θ_s = 14°, and β_1 = 77° ($q_{1\parallel}$ = 0.4 a.u.). Positive values of $q_{2\parallel}$ indicate $\mathbf{q}_2$ parallel to $\mathbf{q}_1$, negative values indicate $\mathbf{q}_2$ antiparallel to $\mathbf{q}_1$. The asymmetry of the cross section about β_2 = 270° indicates a preference for the two electrons to travel in the same direction under the specified conditions. Cross sections and probability densities were calculated with a correlated CI wave function.

To measure two-electron momentum densities the present tandem spherical analyzer will be retained for scattered electron analysis, however, two toroidal electrostatic analyzers will replace the single truncated spherical analyzer of the present experiment. The analyzers will be positioned to collect electrons ejected into planes that are rotated about the scattered direction by +60° and –60° relative to the plane containing $\mathbf{k}_0$, $\mathbf{k}_s$, and $\mathbf{K}$. An array of up to 8 electron multipliers positioned at the image plane of each analyzer will accept a range of ejected-electron angles about the directions $\mathbf{K}_1$ and $\mathbf{K}_2$ to be studied as well as increase the data rate by a factor of 64. Preliminary measurements will be made at coarse momentum resolution using existing electron detectors. Final measurements will be performed with an array of electron multipliers custom designed to achieve a momentum resolution of ±0.1 a.u. These detectors will cover a range from 0 to 1 a.u. with a resolution of ±0.1 a.u. The pass energies and energy resolutions of the analyzers are fixed by the need to maintain impulsive collision conditions while at the same time defining the final state of the residual doubly charged ion and excluding Auger electrons.

For magnesium, the large separation in the energies of the ion states means that it will be possible to measure two-electron momentum densities for the ejection of two 3s electrons while unambiguously establishing the final ion state.

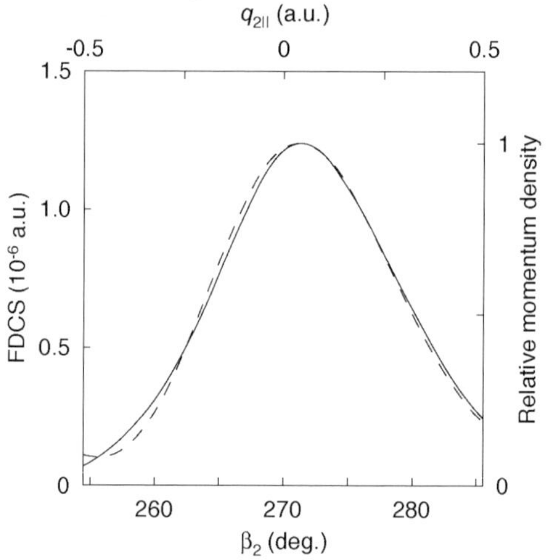

Figure 6. Mg $3s^2$ two-electron momentum probability density (solid line). (e,3e) five-fold differential cross section (FDCS) (dashed line).

5. Signal rates and momentum resolution

The feasibility of a triple coincidence measurement is determined by data accumulation time required to achieve a desired statistical precision, rather than just the coincidence rate. Statistical precision depends on both the true triple coincidence electrons and the coincidence rate due to randomly correlated electrons (Lahmam-Bennani 1991). Once the random coincidence rate equals the true coincidence rate, data accumulation times required to achieve a desired precision increase with increasing incident current. This limit can be raised by improving coincidence timing resolution, analyzer efficiency, and the uncorrelated background signal.

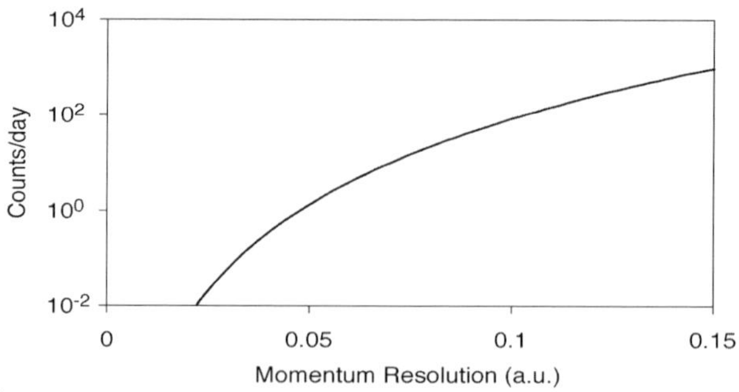

Figure 7. Calculated double ionization signal rate as a function of the experimental momentum resolution.

With readily attainable improvements in these areas a 36-fold increase in the true to random ratio can be achieved over the present spectrometer and incident currents up to 20 μA can be used. A plot of the expected true coincidence rate as a function of momentum resolution is shown in Fig. 7. The coincidence rate was calculated for a true to random ration of 1:1, which is obtained at an incident current of 20 μA. At a momentum resolution of ±0.1 a.u. the maximum true coincidence rate is expected to be 86 counts/day. Table II summarizes the operating parameters of the proposed experiment.

Table II

Parameter	$\Delta q = \pm.0.15$ a.u.	$\Delta q = \pm.0.10$ a.u
Electron Energy Resolution	8.0 eV	8.0 eV
Angular Resolution, Scattered Electrons	±1.1°	±0.7°
Angular Resolution, Ejected Electrons	±4.1°	±2.7°
Signal Rate	978 counts/day	86 counts/day
Signal-to-Random Ratio	1:1	1:1
Incident Current	20 μA	20 μA

6. Conclusions

In order to measure two-electron momentum densities using electron impact double ionization it is necessary to work in a kinematic regime where the ionizing collisions occur impulsively and the double ionization mechanism preserves the information about the momenta of the target electrons at the instant of their ejection. Two impulsive double ionization mechanisms, TS1 and TS2, have been identified from the angular distributions of the ejected electrons about the momentum transfer direction. Of the two mechanisms, only TS2 is suitable for momentum density measurements. TS2 can be isolated from TS1 by changing the angle between the momenta of the ejected electrons and the momentum transfer vector at which the ejected electrons are detected. An instrument based on the results and realistic calculations will be able to directly measure two-electron momentum densities with a resolution of from ±0.15 to ±0.10 a.u. over data accumulation times from 1 to 10 days. It is anticipated that these measurements will offer the first experimental measurement of the correlated motion of electrons in atoms.

7. Acknowledgement

Research supported by NSF Grant No. PHY-99-87870.

8. References

Amusia M Ya 1993 J. Phys. IV **3**, 91.

Coplan M A, Moore J H, Doering J P 1994 Rev. Mod. Phys. **66** 985

Coplan M A, Doering J P, Madison D H, Moore J H, and Pinkas A A 1996 Aust. J. Phys. **49** 321

Dorn A, Kheifets A, Schröter C D, Najjari B, Höhr C, Moshammer R, and Ullrich J 2002 Phys. Rev. A **65** 032709

Dorn A, Kheifets A, Schröter C D, C Höhr, G Sakhelashvili, R Moshammer, J Lower, J Ulrich 2003 Phys. Rev. A **68**, 012715

El-Marji B, Schröter C, Duguet A, Lahmam-Bennani A, Lecas M and Spielberger L 1997 J. Phys. B **30**, 3677

Ezra G S and Berry R S 1983 Phys. Rev. A **28**, 1974

Glassgold A E and Ialongo G 1968 Phys. Rev. **175**, 151

Jia C C, Lahman-Bennani A, Duguet A, Avaldi L, Lecas M, and Dal Cappello C 2002 J. Phys. B **35**, 1103

Jia C.C., Lahmam-Bennani A, Dal Cappello C, Duguet A, and Avaldi L 2003 J. Phys. B **36** L17.

Krause J L and Berry R S 1985 J. Chem. Phys. **83**, 5153

Krause J L, Morgan J D III, and Berry R S 1987 Phys. Rev. A **35**, 3189

Lahmam-Bennani A, Dupré C, and Duguet A 1989 Phys. Rev. Lett. **63**, 1582

Lahmam-Bennani A 1991 J. Phys. B: At. Mol. Opt. Phys. **24**, 2401

Lahmam-Bennani A, Ehrhardt H, Dupré C, and Duguet A 1991 J. Phys. B **24**, 3645

Lahmam-Bennani A, Duguet A, Dal Cappello C, Nebdi H, and Piraux B 2003 Phys. Rev. A **67** 010701 (R)

McCarthy I E and Weigold E 1976 Phys. Rep. **27C**, 275

Pinkás A A, Coplan M A, Jones S, Madison D H, Rasch J, Whelan C T, Allan R J, and Walters H R J in *New Directions in Atomic Physics*, Whelan, Dreizler, Macek, and Walters, eds., Plenum Press, 1998.

Popov Y, Dal Cappello C, Joulakian B, and Kuzmina N M 1994 J. Phys. B **27**, 1599

Smirnov Yu F, Pavlitchenkov A V, Levin V G and Neudatchin V G 1978 J. Phys. B **11**, 3587

Taouil I, Lahmam-Bennani A, Duguet A, and Avaldi L 1998 Phys. Rev. Lett. **81**, 4600

Tweed R J 1992 Z. Phys. D **23**, 309 .

van Boeyen R W, Watanabe N, Doering J P, Moore J H, and Coplan M A 2004 Phys. Rev. Lett. **92**, 223202

Inst. Phys. Conf. Ser. No. 183
Paper presented at Int. Conf. Electron and Photon Impact Ionization and Related Topics, Louvain-la-Neuve, 2004

Electron excitation & ionization of ^{40}Ca at low energy

Andrew James Murray
Schuster Laboratory, University of Manchester, Manchester M13 9PL, United Kingdom

Abstract. Experimental results for excitation and ionization of calcium have been obtained at incident energies from 10 to 65eV. The results for excitation are determined using the super-elastic scattering technique, allowing Atomic Collision Parameters to be extracted. Results for ionization are conducted in a coplanar symmetric geometry, both electrons emerging from the reaction with equal energy. By providing results in this energy regime, it is envisaged models which describe excitation and ionization can be refined to produce a unified theory of the interaction.

1. Introduction

Excitation and ionization of atomic targets by electron impact has been an area of considerable interest since the early days of atomic physics. The dynamics of the reaction are complex, since the interaction between the target and incident electron is driven by the long range Coulomb field. The process is inherently a many body problem for which exact solutions of the Schrödinger equation are not possible, and so it is necessary to invoke approximations when describing the interaction. Models such as the Distorted Wave Born Approximation (DWBA) [1] and the Convergent Close Coupling method (CCC) [2] have had success in different energy regimes, however no single theory has successfully predicted the data over a wide range of impact energies ranging from threshold to high energy. The goal of the experimental work at Manchester is to provide results at low to intermediate energies, a region which spans the regime where different models are successful. It is hoped that this data will then be utilised in the goal to provide a unified theory for electron impact excitation and ionization.

In the experiments described here ^{40}Ca calcium is used as the target. Calcium is an alkaline earth atom and so has a filled $(4s^2)^1S_0$ valence shell outside a closed $(1s^2 2s^2 2p^6 3s^2 3p^6)$ core. The process of electron impact ionization promotes one of the bound electrons to the continuum. To a first approximation, the differential (e,2e) ionization cross section is expected to be similar to that of helium which has a closed $(1s^2)^1S_0$ valence shell. The experimental results however show a marked contrast between the two targets, indicating electrons in the core are playing a significant role in the interaction. Models of the ionization process often invoke a 'frozen core' approximation, however it is clear that this is not valid for calcium, and that the core electrons must be included.

The process of electron impact excitation studied here promotes calcium to the $(4s4p)^1P_1$ state, and the Atomic Collision Parameters (ACP's) defining the alignment and orientation of the excited target are determined using the superelastic scattering technique

[3]. ^{40}Ca calcium is an ideal target to study using this laser-based technique as it has no hyperfine structure. Many of the experimental difficulties associated with the complex hyperfine structure of alkali targets are therefore eliminated [4,5], allowing the experimental data to be measured to high precision for comparison to theory.

A key aim in these studies is to also provide data for ionization from an *excited* target. The method of target preparation used for super-elastic scattering measurements is again used to promote calcium atoms to the $(4s4p)^1P_1$ excited state. In contrast to electron impact ionization from the spherically symmetric $(4s^2)^1S_0$ state, the shape of the initial target state with respect to the scattering geometry is then controlled by the laser prior to ionization. In these experiments the (e,2e) cross section depends both on the scattering geometry and the target preparation, and additional information about the ionization dynamics is obtained. These difficult experiments have presently halted due to lack of funding, however results from initial exploratory experiments are presented.

Finally, experiments are currently being conducted ionizing ^{39}K potassium from the ground state. Potassium has the same closed shell core as calcium, with a single electron in the valence shell. By comparing the ionization cross section from ^{39}K with ^{40}Ca, the effects of correlation between the outer valence electrons is studied. Such comparisons would ideally be conducted between helium and hydrogen, however the difficulty in producing a hydrogen beam of sufficient density precludes this at the present time. The experiments on ^{39}K are in their preliminary stages, however it is already clear a marked difference exists. Examples of results obtained from this target are given in section 6 to promote discussion.

2. The experimental setup.

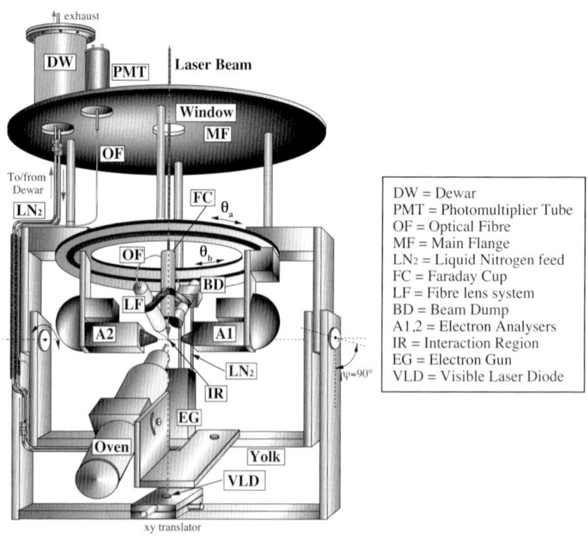

Figure 1. The (e,2e) spectrometer at Manchester configured in the perpendicular plane.

All experiments were conducted using the computer controlled and optimised (e,2e) spectrometer at Manchester. Details of this spectrometer are given in [6], however a number of modifications have been made to allow measurements from calcium and potassium. These changes include the installation of an atomic beam oven in place of the internal photomultiplier tube previously used to align the electron beam, a fibre optic delivery system connected to an external photomultiplier tube used to monitor fluorescence from the interaction region, an internal diode laser to define the direction of the laser beam, a liquid nitrogen cooled beam dump, liquid nitrogen delivery system and external dewar.

The experimental apparatus is depicted in figure 1, configured in the perpendicular plane. The (e,2e) spectrometer can access all geometries from coplanar geometry to the perpendicular plane by rotating the electron gun from $\psi = 0°$ to 90°. All results presented here were in coplanar geometry. The oven was heated to a temperature of 600°C for calcium and 250°C for potassium, whereas the beam dump was maintained at a temperature of –180°C by passing liquid nitrogen through a chamber enclosing the dump. The background pressure during operation was around 7 x 10^{-8} torr when the oven was operating.

A major experimental difficulty working with calcium and potassium is due to deposition of scattered atoms not confined by the beam dump onto sensitive surfaces inside the chamber. Each target produces different problems. Calcium deposition was found to alter the contact potential in the target region, and it was necessary to regularly check the energy of the reaction process to allow for this. Calcium also deposited onto the fibre optic system, rendering this inoperable within 3 days [7]. The effects due to potassium deposition are markedly different. No change is observed in the contact potential during operation, however a severe problem occurs when potassium deposits onto lens elements in the analysers. This renders the electron multipliers inoperable, the electron signal ceases and the spectrometer has to be serviced. At the time of writing this problem has not been solved, although steps to reduce this are being investigated.

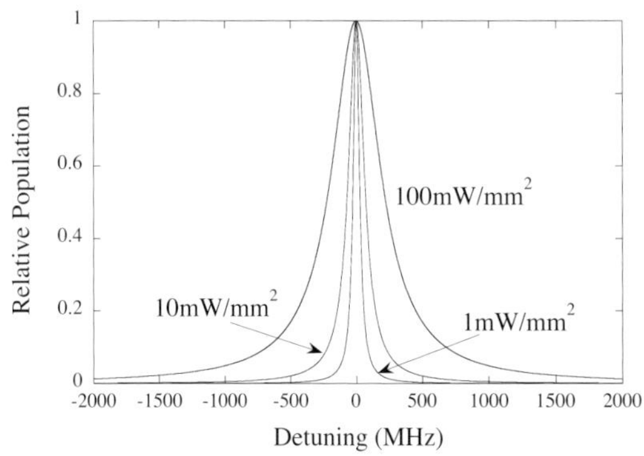

Figure 2. The effect of laser detuning on the population of the 4^1P_1 excited state of calcium.

The laser system used to prepare the atom in an excited state prior to ionization and superelastic measurements is a Coherent MBR-110 Ti:Sapphire laser pumped by a Coherent 10W Verdi laser. Radiation from the MBR-110 laser is frequency doubled by a Coherent MBD-200 doubler, which provides 150mW of radiation at a wavelength around 423nm. The linewidth of the laser radiation is ~1.2 x 10^{-7}nm. The laser system is frequency locked to an external Fabry-Perot etalon within the MBR-110 laser, however the stability of the control system is poor, and the laser drifts up to 2GHz an hour. This frequency shift significantly alters the population density of laser excited targets as shown in figure 2, where the relative population as a function of detuning from resonance is shown for laser intensities ranging from 100mW/mm^2 to 1mW/mm^2. Even at the highest laser intensity, the population reduces to only 1.3% of its initial value for a detuning of ±2GHz. Since the (e,2e) experiment must operate continuously up to 12 hours each day, this drift makes it impossible to carry out detailed (e,2e) ionization measurements at the present time. The superelastic scattering measurements were taken over a much shorter period of time, and so the laser drift was corrected manually as the data accumulated.

3. Excitation studies of calcium using the superelastic technique

Excitation of calcium by electron impact results in an electron scattered through angle θ_e which loses energy equivalent to the excitation energy E_{41P1} of the target state. Both the energy E_{inc} and momentum $\mathbf{k}_0$ of the electron change during the reaction, with the target remaining neutral. Hence there is only one charged particle in the initial and final states of the system. The interaction is conventionally characterised by measuring the scattered electron in coincidence with the photon emitted from the electron excited state:

$$\begin{aligned} e(\mathbf{k}_0;E_{inc}) + Ca(4^1S_0) &\rightarrow Ca^*(4^1P_1) + e(\mathbf{k}_1(\theta_e);E_1 = E_{inc} - E_{4^1P_1}) \\ &\rightarrow Ca(4^1S_0) + \gamma(423nm) + \underbrace{e(\mathbf{k}_1(\theta_e);E_1 = E_{inc} - E_{4^1P_1})}_{\text{detected in coincidence}} \end{aligned} \tag{1}$$

The ACP's describing the 4^1P_1 excited state $P_{lin}, \gamma, L_\perp$ can then be determined as a function of the scattering angle θ_e by measuring the polarization state of the emitted photon in coincidence with the scattered electron [8,9].

By contrast, the super-elastic scattering technique allows the ACP's to be determined without the need for coincidence measurement, by initially exciting the atom using laser radiation, then scattering an electron from the excited target. This reaction is the time reversal of the electron-photon coincidence technique, and only requires the rate of electrons scattered super-elastically from the target to be determined as a function of the scattering angle θ_e and laser polarization [10-12]. The laser is input along the $-z_{nat}$ direction (figure 3), producing an excited target. The incident electron directed at the laser excited target super-elastically scatters with an energy $E_{out} = E_{inc} + E_{41P1}$ and is detected along the x_{nat} axis. The laser polarization is varied using a linear polarizer, λ/2 and λ/4 plates and the rate of superelastically scattered electrons is measured. The signal accumulates thousands of times faster than with coincidence methods, this high rate of data accumulation leading to a corresponding decrease in the uncertainty of the measurements.

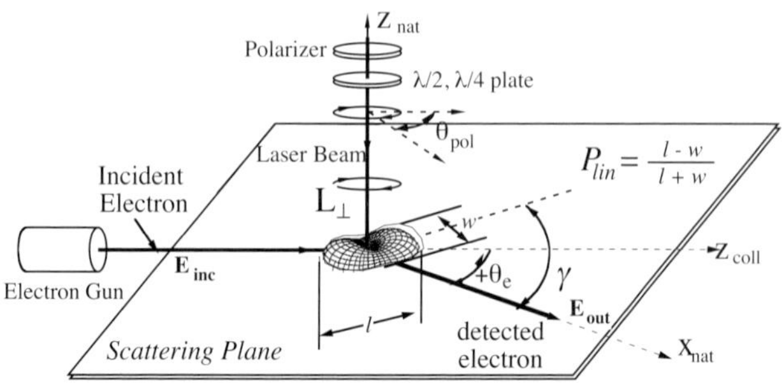

Figure 3. The superelastic scattering geometry showing the excited state with respect to the incident electron beam and scattered electron. The ACP's are as shown.

P_{lin}, γ and $L_{\perp}$ reveal the shape of the target P-state that would be produced by an inelastic electron collision as shown in figure 3. P_{lin} determines the ratio of the length to width of the P-state, whereas γ indicates the direction of the major axis with respect to the x_{nat} axis. $L_{\perp}$ indicates the degree of angular momentum transferred to the target during the collision. It is these parameters that are compared to theory.

The laser pumping process used to produce the excited targets adds a complexity to the reaction which must be determined. Alkali atoms such as sodium and potassium have a rich hyperfine structure, and the ACP's are related to the super-elastic measurements through optical-pumping coupling parameters determined from experiment [4,5]. By contrast, the absence of hyperfine structure in ^{40}Ca ensures a direct relationship exists between the ACP's and measurement. No additional uncertainty accumulates beyond that measured, and the results can be presented with a high degree of confidence [9].

Figure 4 shows the ACP's obtained for the 4^1P_1 state of calcium at incident energies equivalent to 35eV and 20eV for inelastic scattering, compared to the Relativistic Distorted Wave (RDWA) theory of Stauffer and colleagues [13]. The agreement between theory and experiment is striking at the higher energy, indicating the major attributes of the reaction have been included in the model. The main disagreement is found for P_{lin}, where theory predicts values close to unity between $\theta_e = 40$ and 80°, whereas experiment show P_{lin} closer to 0.9. Both γ and $L_{\perp}$ agree well over all angles studied.

At 20eV, differences between theory and experiment are obvious. In particular, the experimental results for P_{lin} show a deep minimum at around $\theta_e = 120°$ not predicted by theory. The results for γ are in reasonable agreement except near $\theta_e = 120°$, whereas the results for $L_{\perp}$ are in poor agreement over all angles. The RDWA model (which is principally a high energy model) is failing at this energy.

These results have been sent to a number of theoretical colleagues who adopt different techniques to describe the interaction. The CCC model of Bray and co-workers [2] has proven to be particularly successful in describing the interaction for alkali atoms, and it will be interesting to see how well this model performs for calcium.

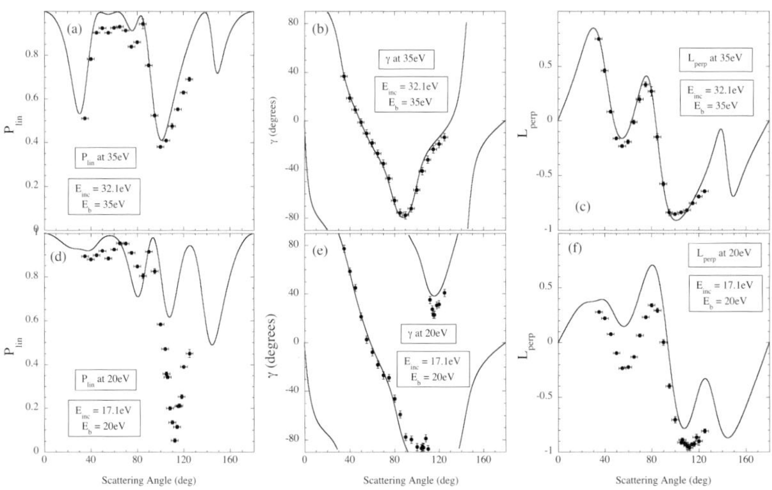

Figure 4. Examples of the ACP's for excitation of calcium to the 4^1P_1 excited state at 20eV & 35eV.

4. Ionization studies of calcium from the 4^1S_0 ground state

Modelling single ionization of a neutral target by electron impact is a considerably more difficult problem than that of excitation, since the reaction produces three charged particles in the final state. The outgoing electrons can share the excess energy from the reaction in different ways, and may emerge from the reaction in any direction. Both the energy (E_a, E_b) and momenta ($\mathbf{k}_a$, $\mathbf{k}_b$) of the resulting electrons depend on the ionization potential (IP), the interaction between the incident electron and the target, the interaction between the electrons and the ionic core and their mutual interaction. The most information extracted from the reaction occurs when the electrons are detected in coincidence, and the angular correlation between the electrons is extracted:

$$e(\mathbf{k}_0;E_{inc}) + Ca\left(4^1S_0\right) \rightarrow Ca^+ + \underbrace{e(\mathbf{k}_a(\theta_a,\phi_a);E_a) + e(\mathbf{k}_b(\theta_b,\phi_b);E_b = E_{inc} - IP - E_a)}_{\text{measured in coincidence}} \quad (2)$$

It is usual in these (e,2e) experiments to restrict the scattering geometry, and in the experiments carried out here the incident, scattered and ejected electrons all lie in the same plane (coplanar geometry). The reaction is further constrained to select electrons emerging from the reaction at equal and opposite angles ($\theta_a = \theta_b = \theta$) with respect to the incident momentum vector, and with equal energy ($E_a = E_b = (E_{inc} - IP)/2$).

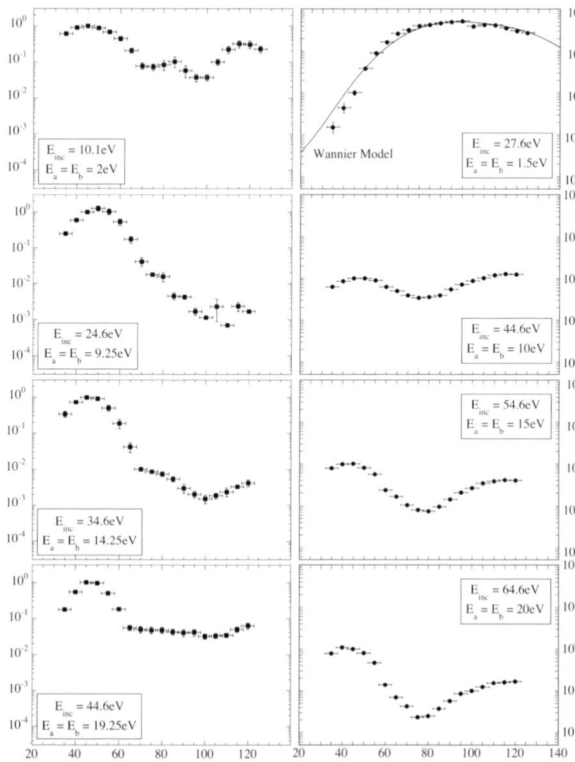

Figure 5. Ionization of calcium compared to helium at selected energies in coplanar symmetric geometry. Note the different scale for helium at 27.6eV incident energy. For details, see text.

Figure 5 shows examples of the result of these studies at selected outgoing electron energies compared to results taken for helium at similar energies. There are at present no theoretical results with which the ^{40}Ca measurements can be compared. The results show a marked difference between ionization of helium and calcium. At the lowest energy ($E_a \sim 2$eV) helium ionization is well described using the Wannier model [14]. In this case the electrons emerge predominantly opposite each other ($\theta = 90°$) due to the post-collisional Coulomb interaction between the electrons. For calcium there is some evidence of structure at $\theta = 90°$, however the probability of ionization is significantly higher in both the forward and backward scattering regions. At an energy $E_a \sim 10$eV, ionization of helium is equally probable in the forward and backward regions, whereas forward scattering dominates the ionization of calcium by a factor of 1000 : 1. At $E_a \sim 20$eV forward scattering is dominating both helium and calcium cross sections, however the structure of the angular correlation is quite different. Helium shows a minimum at $\theta \sim 75°$, whereas the cross section for calcium shows no dip, and flattens off from 60° to higher scattering angles.

From these results it is clear that the ionization of calcium does not emulate that of helium, and that the effects of the core electrons is important over all energies studied. The effects of post-collisional interactions must become important at energies approaching threshold for calcium, however the energy where this becomes significant is clearly lower than for helium. At higher energies the forward scattering peak in helium can be explained as due to a binary collision between incident and target electrons, whereas the backward scattering peak can be explained by elastic scattering of the incident electron from the target followed by a binary collision. This simple description cannot be applied to calcium, which shows different structure depending on both energy and scattering angle. It remains to be seen how successful models of the interaction process will be to predict these effects.

5. Ionization studies of calcium from the 4^1P_1 excited state

A major challenge to both theory and experiment is the study of ionization from an excited target. An excited atom with $J \neq 0$ can be described by a coherent linear summation over the m_J basis states, leading to alignment and orientation of the target prior to the reaction. For calcium the excited 4^1P_1 state created by laser excitation can be described as

$$|\psi\rangle = \sum_{m_J=-1}^{+1} a_{m_J} |J=1,m_J\rangle = a_{+1}|1,1\rangle + a_0|1,0\rangle + a_{-1}|1,-1\rangle \qquad (3)$$

where the basis states are here defined in the collision frame of the experiment. The complex amplitudes a_{mJ} define the magnitude and relative phase between the states required to describe the excited state. Ionization from the excited state depends on the interaction with respect to the chosen basis set, since the target amplitudes a_{mJ} vary as the laser polarization and direction is changed.

Figure 6 shows an example of the process under study for the 4^1P_1 state of calcium. In this example the laser produces a fully aligned atom with major axis either along the incident electron direction, or perpendicular to this direction. Choosing a quantization axis along the incident beam therefore produces initial target states given by

$$|\psi\rangle_{\parallel} = |1,0\rangle; \qquad |\psi\rangle_{\perp} = \frac{1}{\sqrt{2}}\left(|1,+1\rangle - |1,-1\rangle\right) \qquad (4)$$

Ionization from each laser aligned target will depend upon the strength of the interaction from the contributing basis states. The (e,2e) differential cross section now depends upon a *coherent* summation over these states, in contrast to ionization from an atom incoherently prepared (eg from a thermally produced beam of atoms in a ground P-state).

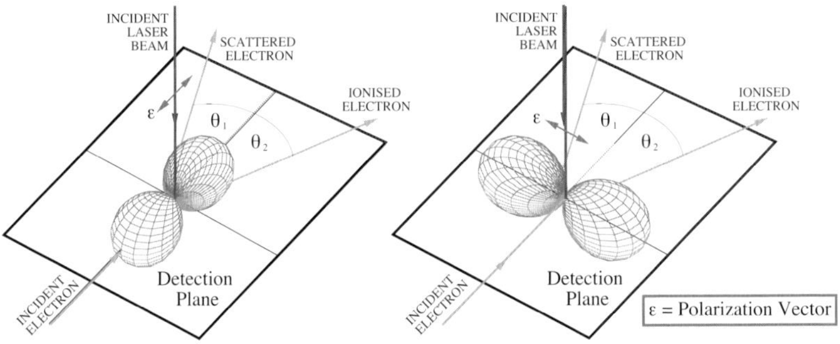

Figure 6. The (e,2e) geometry from laser excited calcium showing the geometry for the state aligned parallel and perpendicular to the incident electron beam.

It is the goal of experiments on laser excited atoms to test the influence of this state preparation on the measurement. Since the reaction process is controlled by the laser, additional information on the dynamics is obtained. Initial experiments have been conducted using calcium prepared as in figure 6, for an incident energy of 34eV at a scattering angle $\theta = 35°$. The coincidence signal obtained from the ground (a) and excited states (b,c) is shown in figure 7, for the laser de-tuned from resonance (b) and tuned to resonance (c). The number of excited atoms which are produced is calculated to be ~15% of the total population, and this is reflected in the lower coincidence count measured. A ratio of the cross section σ measured when the laser is parallel and perpendicular to the incident electron beam is $\sigma_{\parallel}/\sigma_{\perp} = 1.35 \pm 0.25$. It is clearly desirable to make more comprehensive measurements for a range of laser directions, polarizations, scattering geometries and electron energies to explore the influence of the initial target state on the ionization process, and it is hoped that these experiments can re-commence in the near future.

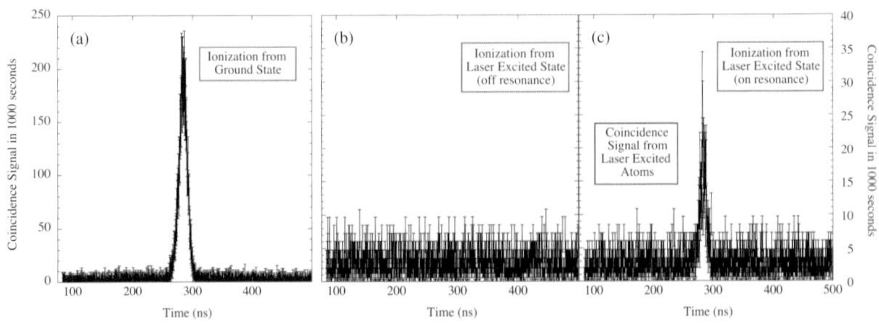

Figure 7. Coincidence signals ionizing from (a) the ground state, (b) the excited state with the laser de-tuned and (c) the excited state with laser tuned to resonance.

6. Ionization of potassium from the $4^2S_{1/2}$ ground state

(e,2e) measurements have now commenced from the $4^2S_{1/2}$ ground state of potassium, in a coplanar symmetric geometry at incident energies spanning the range where data on calcium was obtained. The difference found between measurements using calcium and potassium test the effects of correlation between the valence electrons, since the core is identical for both targets. These measurements have only recently started, and results at an incident energy of ~24eV and ~34eV are shown in figure 8. In contrast to calcium at these energies, the ratio between forward and backward scattering is now significantly larger at the lower energy than at the higher energy. This is in direct opposition to measurements in other closed shell targets in this energy regime [15,16]. It is too early to draw definitive conclusions from these results, but it appears that ionization of potassium is showing significant differences to calcium, although similarities exist in the overall shape of the cross section. These experiments are continuing, and it remains to be seen how the results develop as the energy is varied from threshold through to higher energies.

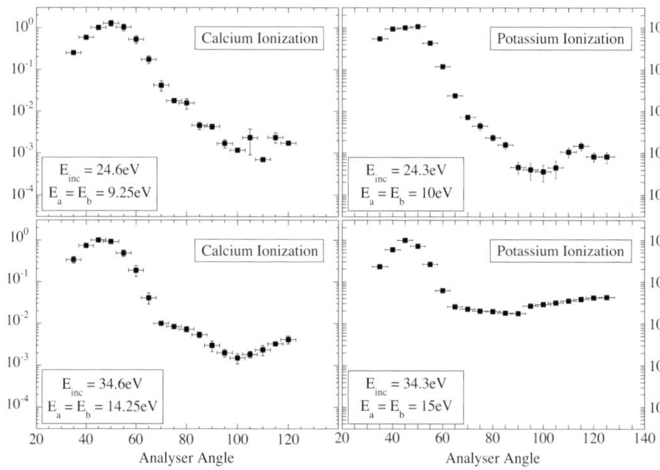

Figure 8. Preliminary results comparing potassium and calcium ionization at E_{inc} ~24eV and ~34eV.

7. Conclusions and future work.

Study of ionization and excitation of calcium in the low to intermediate energy regime has produced a comprehensive set of data with which to compare to theory. Calculations based on a relativistic distorted wave approximation produce excellent agreement for excitation at higher energies, however this theory does not agree with experiment at lower energies. No theoretical calculations have yet been conducted for ionization, and it will be interesting to see how successful the different models are. Ionization experiments have also been conducted from the excited state of calcium. These are in their infancy, so it remains to be seen what the effects of varying the initial target state will be. Finally, experiments ionizing potassium have commenced, which will elucidate the effects of electron correlations in the outer valence shell when compared to the results from calcium.

New experiments are soon to start which will probe the excitation of calcium and potassium over the full range of scattering angles from 0° to 180° using the superelastic scattering method. These experiments will utilise a new type of Magnetic Angle

Changing (MAC) spectrometer invented in Manchester [17] coupled with a high resolution low energy electron gun. In this way the excitation process will be explored over the complete angular range from threshold for the first time.

It is the aim of the work at Manchester to provide data over a wide range of scattering geometries in an energy regime which spans the regions where different models are successful. In this way, low energy models can be refined for higher energies, and high energy models can be modified for lower energies. It is hoped that through this process a unified model of the scattering process will eventuate. The results presented here go part of the way to achieving this goal.

Acknowledgements

We would like to thank the EPSRC in the UK for providing funds to carry out these studies. We would also like to thank the technicians in the mechanical and electronic workshops at the Schuster Laboratory for providing excellent technical support.

References

[1] Walters H R J, Zhang X & Whelan C T 1993 *(e, 2e) and Related Processes* (Dordrecht: Kluwer) p33

[2] Bray I, Fursa D V, Kheifets A S and Stelbovics A T 2002 *J. Phys. B* **35** R117

[3] Hertel I V and Stoll W 1977 *Adv. At. Mol. Phys.* **13** 113

[4] Farrell P M, MacGillivray W R and Standage M C 1991 *Phys. Rev.* A **44** 1828

[5] Farrell P, MacGillivray W R and Standage M C 1988 *Phys. Rev.* A **37** 4240

[6] Murray A J, Turton B C H and Read F H 1992 *Rev. Sci. Instrum.* **63** 3346

[7] Murray A J & Cvejanovic D 2003 *J. Phys. B* **36** 4875

[8] Andersen N, Gallagher J W and Hertel I V 1988 *Phys. Rep.* **165** 1

[9] Murray A J & Cvejanovic D 2003 *J. Phys. B* **36** 4889

[10] Karaganov V, Bray I, Teubner P J O and Farrell P 1996 *Phys. Rev.* A **54** R9

[11] Shurgalin M, Murray A J, MacGillivray WR, Standage MC,Madison D H,Winkler K D and Bray I 1998 *Phys. Rev. Lett.* **81** 4604

[12] Hall B, Shen Y, Murray AJ, Standage MC, MacGillivray WR & Bray I 2004 *J. Phys. B* **37** 1113

[13] Srivastava R, Zuo T, McEachran R P and Stauffer A D 1992 *J. Phys. B* **25** 3709

[14] Wannier G H 1953 *Phys. Rev.* **90** 817

[15] Murray A J and Read F H 2000 *J. Phys. B* **33** L297

[16] Murray A J, Bowring N J and Read F H 2000 *J. Phys. B* **33** 2859

[17] Cubric D, Mercer, D J L Channing J M, King G C K and Read F H 1999 J Phys B **32** L45

Inst. Phys. Conf. Ser. No. 183
Paper presented at Int. Conf. Electron and Photon Impact Ionization and Related Topics, Louvain-la-Neuve, 2004

Role of the ground state in (e,3e) processes on helium

L.U. Ancarani, T. Montagnese and C. Dal Cappello

L.P.M.C., Institut de Physique, Technopôle 2000, 57078 Metz, France

Abstract. The role played by the ground state wave function in (e,3e) processes on helium is investigated by comparing calculations with absolute measurements performed at high incident energy. We first address the question of the validity of the first Born approximation (FBA) and then critically analyze the FBA results which successfully reproduced these experimental data [Phys. Rev. Lett. **91**, 73201 (2003)]. It has been suggested that the good agreement is strongly related to the quality of the initial helium ground state (Pluvinage wave function), in particular to the fact that it is diagonal in all Coulomb interactions. Here, we show that this conclusion is not correct since the same agreement can be obtained with a new wave function that does not diagonalize the hamiltonian. Finally, a simulation which takes into account the uncertainty on the scattering angle and, somehow, the energy resolution of the ejected electrons is presented.

1. Introduction

The study of kinematically complete double ionization experiments by electron impact is a very powerful tool to investigate the role of electron-electron correlation. The so-called (e,3e) experiments in which the three electrons (the two ejected and the scattered) are detected in coincidence allow one to check with a good accuracy the different theories (see the recent review paper [1]).

In this contribution we shall investigate the role played by the ground state wave function in the calculation of cross sections for processes (e,3e) on helium. This will be done by comparing the results with the absolute cross sections (accuracy of 25%) measured in the coplanar (e,3e) experiments of Lahmam-Bennani *et al.* [2], performed at an incident energy of $\sim$5.6 keV and under a small projectile's scattering angle of 0.45^0 (momentum transfer of 0.24 a.u.; dipolar regime). The two ejected electrons have been detected with equal energy (10 eV).

Several theoretical calculations have yielded contrasting results which shall be briefly discussed in section 2. For the experimental kinematics of [2] the first Born approximation (FBA) - with respect to the interaction of the fast projectile with the target - should apply [3]. Moreover, the 3C model [4, 5] - where the final state is described by a product of three Coulomb waves - seems also sufficient. Within this model, remains the question of what to use for the description of the helium bound state. Hydrogenic and Hylleraas wave functions

[6] yield cross sections similar in shape and magnitude, but are in bad agreement [3] with the experimental results. Jones and Madison [3] have therefore suggested using the wave function introduced by Pluvinage [7]. The calculated (e,3e) cross sections are then in overall good agreement (in 16 out of the 20 geometrical situations) with the experimental data of [2]. The authors therefore conclude that this agreement is related to the quality of Pluvinage wave function, in particular because it diagonalizes the Hamiltonian in all three Coulomb singularities. In this contribution, we would like to show that this claim is not correct.

In section 2, we briefly describe the 3C model within the FBA and the Second Born Approximation (SBA). The choice and properties of several bound state helium wave functions are discussed in section 3. Calculated cross sections with these initial state wave functions are compared in section 4. Concluding remarks are given in section 5.

2. 3C model for (e,3e) cross sections

The five-fold differential cross section (FDCS) corresponding to (e,3e) processes on helium is given by

$$\frac{d^5\sigma}{d\Omega_a d\Omega_b d\Omega_s dE_a dE_b} = \frac{k_s k_a k_b}{k_i} |T_{fi}|^2 \qquad (1)$$

where $d\Omega_s, d\Omega_a$ and $d\Omega_b$ denote, respectively, the elements of solid angle for the scattered and the two ejected electrons. The energy intervals of the two ejected electrons are represented by dE_a and dE_b. The momenta of incident, scattered, the first and second ejected electrons are denoted, respectively, by k_i, k_s, k_a and k_b.

The dominant theoretical models used to describe (e,3e) processes are [1]

- the 3C model (or BBK model) [4, 5] where the final state is described by a product of three Coulomb waves (asymptotically exact double continuum) which account for each two-body interaction (between each electron and the nucleus, and the electron-electron repulsion).

- the Convergent Close Coupling (CCC) model [8, 9]. This model consists in solving the Lippmann-Schwinger equation for the three-body problem and in identifying the double ionization process as an excitation of the positive-energy pseudo-state of the ion.

- the Green function expansion of [10] which corresponds to the well established Faddeev equations for three particles (it is the lowest order iteration of an incremental approach to the four-body Coulomb problem).

- the 6C model [11, 12, 3] corresponds to a product of six Coulomb waves and takes into account the pairwise final state interactions between all four particles (the scattered electron, the two ejected electrons and the nucleus). Note that, this model implicitly includes some contributions from all higher terms of the Born series.

The Green function expansion [10] with Slater type Helium wave function gives reasonable good agreement (shape and magnitude) in 6 out of the 20 experimental situations [2]. The CCC model [9] yields agreement with the shapes of the cross sections but it is not able to reproduce its magnitude (factor 3 too low); note, though, that this model gives good agreement with the relative experimental data of Dorn *et al* [13]. As regards the description of the final state, the 3C model seems sufficient since the results are seen to be very close to those obtained with the 6C model [3] (this, however, does not imply the validity of the FBA).

In this contribution we shall consider the 3C model only, within the FBA and the SBA. The matrix element can be written as the sum of a first and second Born terms

$$T_{fi} = T_{fi}^{B1} + T_{fi}^{B2}. \tag{2}$$

When considering high impact energies, it is reasonable to describe the incoming and scattered electrons by plane waves so that the FBA matrix element reads [1, 14]

$$T_{fi}^{B1} = -\frac{1}{2\pi} < e^{i\vec{k}_s\vec{r}_0}\Psi_f^-(\vec{r}_1,\vec{r}_2,\vec{k}_a,\vec{k}_b)|V(r_0,r_1,r_2)|\Psi_i(\vec{r}_1,\vec{r}_2)e^{i\vec{k}_i\vec{r}_0} > \tag{3}$$

where V is the Coulomb interaction between the projectile and the helium atom ($Z = 2$),

$$V(r_0,r_1,r_2) = -\frac{Z}{r_0} + \frac{1}{|\vec{r}_0 - \vec{r}_1|} + \frac{1}{|\vec{r}_0 - \vec{r}_2|}, \tag{4}$$

(r_0 is the distance between the incident electron and the nucleus, r_1, r_2 are the distances between one of the helium electron and its nucleus). In eq. (3), Ψ_i is the initial bound state of helium, and exchange between the fast scattered electron and the slow ejected electrons has been neglected. The final state Ψ_f^- in the 3C model is described by a symmetrized product of three coulombic distortion factors [4, 5]. The initial and final states which are not exact eigenfunctions of the same Hamiltonian, are artificially orthogonalized. The calculation of the nine-dimensional integral (3) is reduced to (see [14] for details) (i) a double numerical integration when the initial state Ψ_i is described by a Hylleraas-type function; (ii) a three-dimensional numerical (Gauss-Legendre) quadrature – instead of a six-dimensional numerical quadrature as in [3] - when the Pluvinage wave function is used.

For the SBA, we take the term [15]

$$T_{fi}^{B2} = -\frac{2}{\pi^2}\int \frac{d\vec{k}_t}{K_i^2 K_s^2 (k_i^2 - k_t^2 - 2I_n)}$$
$$< \Psi_f^-(\vec{r}_1,\vec{r}_2,\vec{k}_a,\vec{k}_b)|(-Z + e^{i\vec{K}_i\vec{r}_1} + e^{i\vec{K}_i\vec{r}_2})(-Z + e^{i\vec{K}_s\vec{r}_1} + e^{i\vec{K}_s\vec{r}_2})|\Psi_i(\vec{r}_1,\vec{r}_2) > \tag{5}$$

where $\vec{K}_i = \vec{k}_i - \vec{k}_t$ and $\vec{K}_s = \vec{k}_t - \vec{k}_s$ so that $\vec{K}_i + \vec{K}_s = \vec{k}_i - \vec{k}_s = \vec{K}$ is the momentum transfer. Here we use the closure approximation instead of adding all the intermediate states of the target. We have observed little sensitivity to I_n, treated as a parameter, and we have taken $I_n = 20$ eV in the present calculations.

We have calculated, within the FBA and the SBA, the FDCS in the 3C model and with the Bonham and Kohl (number 14) helium wave function [17]. For illustration purposes, the results (in a.u.) are plotted in Fig. 1 for only one of the 20 geometrical situations ($\theta_a = 319^0$) as a function of the angle of one of the ejected electrons θ_b (positive angles are measured counterclockwise, starting from the incident beam direction). The second born term is seen to play a minor role (similar results are obtained at other experimental situations of [2]).

For the high incident energy of the measurements of [2], we expect the FBA to be sufficient, though the calculations of Berakdar [10], with the simplest Slater helium wave function, seem to indicate otherwise. The validity of the FBA is supported by the fact that the SBA

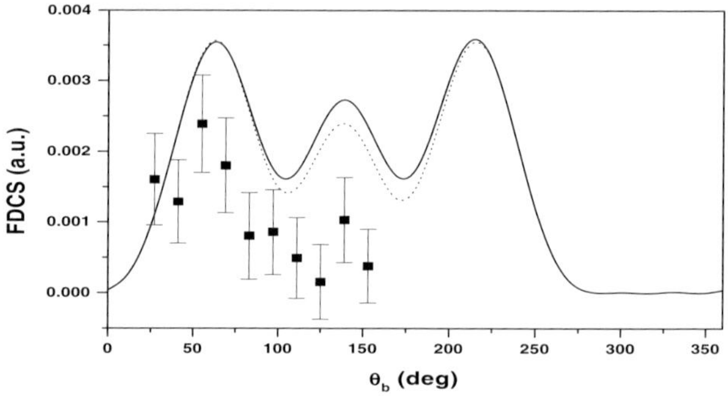

Figure 1. Five-fold differential cross section (FDCS) for (e,3e) ionization of the helium ground state, as a function of the angle of one of the ejected electrons θ_b. The other ejected electron is detected at $\theta_a = 319^0$. The incident electron (at $\sim$ 5.6 keV) is scattered at 0.45^0, and the two ejected electrons escape with equal energy (10 eV). The absolute experimental data [2]: full squares. The two curves are obtained with the 3C model with the Bonham and Kohl [17] - number 14 - initial wave function for the helium ground state: FBA (solid line) and SBA (dotted line).

results are not substantially different from those in the FBA: (i) in the CCC model with a multiconfiguration Hartree-Fock helium wave function [16]; (ii) in the 3C model with the Bonham and Kohl helium wave function (see above). The validity of the FBA is further supported by the fact that little difference is seen between the 6C calculations (with Pluvinage wave function) for electron and positron impact [3].

3. Helium ground state wave function

Jones and Madison [3] have claimed that the good agreement of their theoretical calculations (3C model within the FBA) with experimental (e,3e) data, is directly related to the quality of the Pluvinage wave function for the helium bound state, in particular because it diagonalizes the Hamiltonian in all three Coulomb singularities. In this section, we shall analyze some properties of this and other initial state wave functions to check whether this is true.

Let us start with the non-relativistic Schrödinger equation (SE) for the helium atom ($Z = 2$), in which we neglect the nucleus mass effects. Let r_1 and r_2 be the two electrons radial coordinate, and $r_{12} = |\vec{r}_1 - \vec{r}_2|$ the electron-electron distance. Considering the $^1S_0$ ground state only, the six-dimensional SE reduces to the well known three-dimensional Hylleraas equation (HE) which can be expressed as

$$H\Psi(r_1, r_2, r_{12}) = E\Psi(r_1, r_2, r_{12}) \tag{6}$$

where the Hamiltonian H can be written as

$$H = H_1 + H_2 + H_{12} + H_{12}^{'} \tag{7}$$

with

$$H_i \quad = -\frac{1}{2}\left(\frac{\partial^2}{\partial r_i^2} + \frac{2}{r_i}\frac{\partial}{\partial r_i}\right) - \frac{Z}{r_i} \qquad (i = 1, 2) \tag{8}$$

$$H_{12} = -(\frac{\partial^2}{\partial r_{12}^2} + \frac{2}{r_{12}}\frac{\partial}{\partial r_{12}}) + \frac{1}{r_{12}} \tag{9}$$

$$H'_{12} = -\frac{r_1^2 - r_2^2 + r_{12}^2}{2r_1 r_{12}}\frac{\partial^2}{\partial r_1 \partial r_{12}} - \frac{r_2^2 - r_1^2 + r_{12}^2}{2r_2 r_{12}}\frac{\partial^2}{\partial r_2 \partial r_{12}}. \tag{10}$$

No analytical solution exists today for eq. (6). Note that three singularities (i.e., $r_1 = 0, r_2 = 0$ or $r_{12} = 0$) appear in H, and the factors multiplying the mixed partial derivatives in H'_{12} are always finite. The mean energy

$$E = \frac{<\Psi|H|\Psi>}{<\Psi|\Psi>} \tag{11}$$

is usually minimized with trial wave functions (variational method). The energy, which can be considered as *numerically exact*, of the ground state of (6) is $E_{exact} = -2.903724$ a.u. [18].

Let us define the local energy [19]

$$E(r_1, r_2, r_{12}) = \frac{1}{\Psi(r_1, r_2, r_{12})} H\Psi(r_1, r_2, r_{12}). \tag{12}$$

Only for the exact solution, this local energy will be a constant throughout the entire configuration space, i.e. an eigenvalue of (6), and hence equal to the mean energy (this is the case, for example, for the hydrogenic ions of nuclear charge Z, for which the local and mean energies are equal to $E = -Z^2/2n^2$ (a.u.)). For any other trial wave function, the local energy will not be a constant value; the averaging over the whole space (see eq. (11)) smooths out the variations of the local energy and provides a finite constant value as close as possible to the numerically exact value. For the study of collisions with a helium atom, it is necessary to have a good wave function and not only a good mean energy (the latter is enough for spectroscopical purposes). Obtaining a good mean energy is not a sufficient criterion to state that a trial wave function is good [20]. Having a almost constant local energy is a much stronger quality test.

Bartlett [19] h as shown that Hylleraas wave functions provide local energies ranging from $-\infty$ to ∞. Infinite local energies can come from two sources: (i) the wave function has nodes and $H\Psi \neq 0$; (ii) one or several of the singularities of the potential terms in H_1, H_2 and H_{12} (i.e. $r_1 = 0, r_2 = 0$ or $r_{12} = 0$) are not compensated by the corresponding kinetic terms. The study of the singularities has led Kato [21] to provide mathematical conditions that Ψ must satisfy (the so-called "cusp conditions") in order to eliminate these singularities.

If the electron-electron interaction is neglected, r_{12} is no longer needed in Ψ, and the exact solution of (6) is the product of two hydrogenic He^+(1s) functions $e^{-Z(r_1+r_2)}$, with mean or local energy equal to $-Z^2$. When including the electron-electron interaction, Ψ will depend also on r_{12} and the electronic structure of the helium ground state is substantially modified. A first approach is to consider wave functions of the kind

$$f(r_1)f(r_2)h(r_{12}) \tag{13}$$

where $h(r_{12})$ is a correlation function. The product (13), however, can never be the solution of eq. (6) since it is not separable in the three variables r_1, r_2, r_{12} due to the presence of H'_{12}.

For the purpose of the present study, let us ignore the H'_{12} term of the Hamiltonian (this term has no singularity). This corresponds to replacing the three-body system to 3 two-body systems; equation (6) is then fully separable and the solution can be written as in (13). Consider the choice $f(r_i) = e^{-Zr_i}$, and $h(r_{12})$ an eigenfunction of H_{12}. The latter are exact solutions of the well known Coulomb-type differential equation (without centrifugal term, $L=0$) and are given by the Coulomb-type functions [22] with either positive ($\varepsilon > 0$) or negative ($\varepsilon < 0$) eigenvalue [14]. Considering only regular solutions at $r_{12}=0$, we have

$$g_+(r_{12}) = e^{-kr_{12}}\,{}_1F_1(1+\tfrac{1}{2k},2,2kr_{12}) \quad \text{with} \quad \varepsilon = -k^2 < 0 \tag{14}$$

$$g_-(r_{12}) = e^{-ikr_{12}}\,{}_1F_1(1-i\tfrac{1}{2k},2,2ikr_{12}) \quad \text{with} \quad \varepsilon = k^2 > 0 \tag{15}$$

where k is a real number (note that, because of Kummer's relations [22], k and $-k$ give the same solutions). Both $g_+(r_{12})$ and $g_-(r_{12})$ are real functions of r_{12}: $g_+(r_{12})$ corresponds to bound solutions (in the sense of negative energy) while $g_-(r_{12})$ to continuum solutions. Hence, the wave function (13)

$$e^{-Zr_1}e^{-Zr_2}g_\pm(r_{12}) \tag{16}$$

is a product of three Coulomb functions and is analogous to the 3C final state. Each of the two-body Coulomb interactions for the bound state is treated exactly and hence the Kato cusp conditions are automatically satisfied. If we ignore H'_{12}, the local (and mean) energy is given by $E=-Z^2 \mp k^2$. In the case of the bound solutions $g_+(r_{12})$, we have a special situation when $1+1/2k=-N$ where $N>0$ is an integer. Indeed, the confluent hypergeometric function reduces then to a Laguerre polynomial of order N i.e. $L_N^1(kr_{12})/(N+1)$ and $g_+(r_{12})$ to a hydrogenic function (exponential times a polynomial). The use of a correlation function $h(r_{12})$ in the form of an exponential times a polynomial has been used in many trial wave functions (e.g. [6], [17], [23]). However, usually, the parameters appearing in both the exponential and in the polynomial are chosen in order to optimize the mean energy rather than to eliminate the singularity of the potential term $1/r_{12}$, and hence satisfy Kato's cusp condition.

In the next section, we will present (e,3e) cross sections calculated with four bound state wave functions corresponding to four different correlation functions $h(r_{12})$. The first three, are separable solutions of the kind (16) which automatically satisfy Kato cusp conditions. Two of them, Ψ_0 and Ψ_1, are $g_+(r_{12})$ functions corresponding respectively to $N=0$ ($k_0=-1/2$) and $N=1$ ($k_1=-1/4$). A third one, Ψ_P, is a $g_-(r_{12})$ function corresponding to the value $k=k_P=0.410$ which minimizes the mean energy (this is Pluvinage choice [7] as used by Jones and Madison [3]). The fourth one, Ψ_A, has a $h(r_{12})$ function which is not an eigenfunction of H_{12}. The normalized wave functions, with the corresponding mean energies are,

$$\Psi_0 = N_0\, e^{-Z(r_1+r_2)}\, e^{\frac{1}{2}r_{12}} \qquad E_0 = -2.8561 \text{ a.u.} \tag{17}$$

$$\Psi_1 = N_1\, e^{-Z(r_1+r_2)}\, e^{\frac{1}{4}r_{12}}(1+\frac{r_{12}}{4}) \qquad E_1 = -2.8721 \text{ a.u.} \tag{18}$$

$$\Psi_P = N_P\, e^{-Z(r_1+r_2)}\, e^{-ikr_{12}}\,{}_1F_1(1-i\frac{1}{2k},2,2ikr_{12}) \qquad E_P = -2.8788 \text{ a.u.} \tag{19}$$

$$\Psi_A = N_A\, e^{-Z(r_1+r_2)}\, (1+e^{\frac{1}{4}r_{12}}\frac{r_{12}}{4}) \qquad E_A = -2.8766 \text{ a.u.} \tag{20}$$

where $N_0 = 1.343, N_1 = 1.429, N_P = 1.535$ and $N_A = 1.797$. These four wave functions have comparable mean energies which are not so good when compared to the values obtained with other wave functions available in the literature. In the best case, the Ψ_P wave function yields only about 40 percent of the correlation energy. However, their analytical forms are simple, and practical for collision calculations.

Let us now describe some of the properties and compare the merits of the first three functions Ψ_0, Ψ_1 and Ψ_P which are exact separable solutions. The corresponding correlation functions $h(r_{12})$ are regular at $r_{12}=0$, but behave quite differently at asymptotic distances ($r_{12}\to\infty$) (note that none of the three wave functions has the correct global three-body asymptotic behaviour [24]). For Ψ_0 and Ψ_1, $h(r_{12})$ increases at large distances while, for Ψ_P, it tends to zero in a oscillatory way [14]. This behaviour would seemingly favor the choice towards Ψ_P. One must not forget, however, that the exponential terms $e^{-Z(r_1+r_2)}$ will dominate the large distance behaviour ensuring that the function (16) will represent a bound state. The correlation function $h(r_{12})$ represents the influence of the electron-electron interaction, but does not necessarily represent a physical bound state on its own (the helium three-body system cannot be reduced to the product of 3 two-body systems), and hence does not need to go to zero at large distances.

Since the electron-electron interaction is repulsive it makes sense, at a first glance, to believe that a continuum wave function $g_-(r_{12})$ for $h(r_{12})$ is a better choice than a bound solution $g_+(r_{12})$. Indeed, the former contributes with a positive (rather than negative) energy, and hence provides a local energy ($E=-Z^2\mp k^2$) closer to the exact value. This, however, will not be necessarily true when the full Hamiltonian is considered since the operator $H_{12}^{'}$ mixes the variables r_1, r_2 and r_{12}, and the local energy is not constant anymore.

One of the arguments put forward in the paper of Jones and Madison [3] is that Ψ_P diagonalizes the Hamiltonian in all Coulomb interactions. By construction, both Ψ_0 and Ψ_1 have the same property. As stated in [3], the remaining term in the Hamiltonian, $H_{12}^{'}$, does not have singularities. However, since the Pluvinage wave function Ψ_P has an infinite number of nodes (large distance oscillatory nature), the contribution of $H_{12}^{'}$ to the local energy (12) has an infinite number of infinite values. This is exactly one of the deficiencies of Hylleraas wave functions as pointed out by Bartl ett [19] and quoted by [3]. The wave functions Ψ_0 and Ψ_1, on the contrary, do not have any nodes and provide a local energy finite everywhere [14].

In summary, when comparing the properties of the three wave functions (17)-(19), Ψ_0 and Ψ_1 should not be inferior to the Pluvinage wave function Ψ_P, except possibly for the value of the mean energy.

For the fourth wave function, Ψ_A, given by (20), the corresponding correlation function has no nodes and behaves very similarly to that of Ψ_1. The function Ψ_A is not a separable solution, so that it does not diagonalize the Hamiltonian in all Coulomb singularities and does not satisfy Kato's cusp condition at $r_{12}=0$. In this respect, when compared to the previous three wave functions, Ψ_A should be a much worse representation of the helium ground state.

Some of the properties and deficiencies of the four wave functions can be illustrated by comparing the corresponding local energies (12), as shown in [14]. The variation with respect to the exact numerical value may be represented by the quantity $\delta E=E(r_1,r_2,r_{12})-E_{exact}$: small values of δE are a sign that a wave function behaves well locally. For example, it is easy to see whether the trial wave function satisfies the Kato cusp conditions (which is not the case of Ψ_A), or whether it presents divergencies in the local energy (which is the case of Ψ_P). The comparison of the four wave functions (17)-(20) indicates that Ψ_1 yields a local energy which is overall better than Ψ_0, Ψ_P and Ψ_A, although Ψ_1 produces a slightly worse mean energy.

4. (e,3e) five-fold differential cross sections

The five-fold differential cross sections presented hereafter have been calculated with the 3C model in the FBA, and are compared to the absolute measurements of [2]. For illustration purposes, we have selected two ejected angles θ_a, out of the 20 presented in [2] (calculations

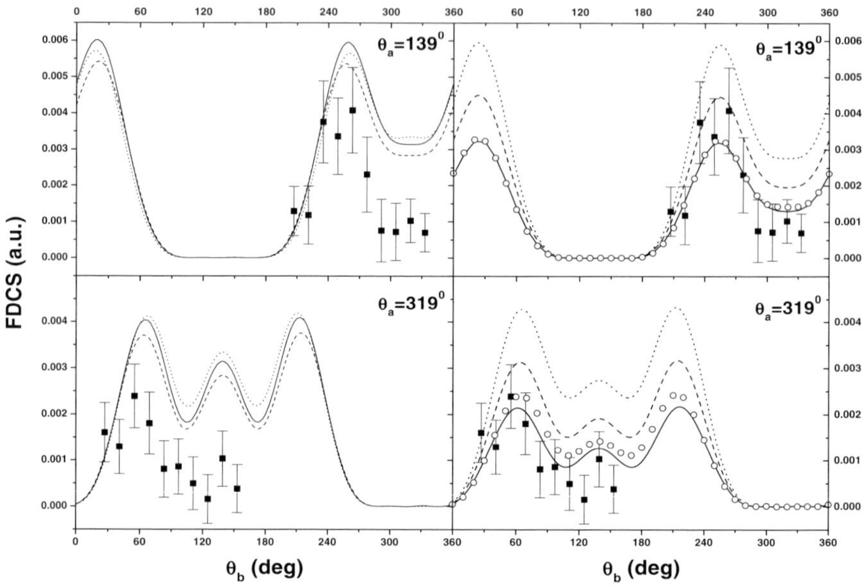

Figure 2. Same as Figure 1 for two angles ($\theta_a = 139^0$ and $\theta_a = 319^0$) indicated in each box. The cross sections have been calculated with different initial wave functions for the helium ground state. Left panels: Bonham and Kohl - number 7 [17] (dotted line), Bonham and Kohl - number 19 [17] (dashed line) and Le Sech [23] (solid line). Right panels: Ψ_0 (dotted line), Ψ_1 (dashed line), Ψ_P (solid line) and Ψ_A (open circles).

at other ejected angles have been done and yield the same conclusions). They correspond to the direction of the momentum transfer $\theta_a = 319^0$ and its opposite $\theta_a = 139^0$. The cross sections are plotted, for a given θ_a, as a function of the angle of the other ejected electron (θ_b).

In the left panels of Fig. 2 we compare the cross sections obtained with three Hylleraas-type initial wave functions for the helium ground state: Bonham and Kohl - number 7 [17] ($E = -2.8757$ a.u.; with radial correlation only), Bonham and Kohl - number 19 [17] ($E = -2.9035$ a.u.; with radial and angular correlations) and Le Sech [23] ($E = -2.9020$ a.u.). The results obtained with these wave functions yield similar shapes and magnitudes, the curves being bunched together. The magnitude is about a factor 2 larger and thus clearly in disagreement with experimental data. As noted in [3], this is quite different from the factor 10 found in [2] with a three-parameter Hylleraas wave function.

In the right panels, we show the FDCS obtained with Ψ_0, Ψ_1, Ψ_P and Ψ_A. The Ψ_P results coincide with those published by [3] (note, though, that their six-dimensional numerical quadrature differs from our three-dimensional quadrature [14]). The shapes are essentially similar to those of the left panels, but the magnitude is lowered by a factor 2, thus giving the good agreement with experimental data found in 16 out of the 20 geometrical situations [3].

The results obtained with Ψ_0 give a similar disagreement in magnitude with the experimental data as those shown on the left panels. On the other hand, the use of Ψ_1 yields cross sections, similar in shape, but with a magnitude which is intermediate between those obtained with Ψ_0 and Ψ_P. It is clear that, Ψ_P gives good agreement in magnitude while the two wave functions Ψ_0 and Ψ_1 do not. Jones and Madison [3] relate this agreement to the fact that the function Ψ_P describes the region of space - where the two electrons are close together - better than the Hylleraas-type wave function. Indeed, Ψ_P satisfies the cusp condition at $r_{12} = 0$, but so do the wave function proposed by Le Sech [23], and the separable solutions Ψ_0 and Ψ_1. One could then argue that Le Sech wave function does not diagonalize the Hamiltonian in the three Coulomb singularities, but Ψ_0 and Ψ_1 do so by construction. The results obtained with Ψ_0 and Ψ_1 indicate therefore that the agreement with experimental data *cannot* be related to the fact that the initial helium wave function diagonalizes the Hamiltonian.

In Fig. 2, we also observe that there is very little difference between the results obtained with the functions Ψ_A and Ψ_P (this is valid for all 20 experimental situations [2], and also for the measurements of Dorn *et al* [13] at 2 keV incident energy [14]). This comparison therefore clearly demonstrates that agreement with experimental data within the 3C model, can be obtained also with a wave function (Ψ_A) which - in principle - cannot be considered as good as the Pluvinage wave function.

Jones and Madison [3] conclude their letter with the following: *"Consequently, electron-electron correlation is treated precisely (and the cusp conditions of Kato are satisfied exactly) by both our initial and final target wave functions and we found that this is crucial for reproducing the absolute measurements"*. Although we agree that treating electron-electron correlation precisely is crucial, we believe that this conclusion is somewhat too hasty [25]. Indeed, the results presented in Fig. 2 clearly show that (i) a wave function (e.g. Ψ_0 or Ψ_1) which diagonalizes the Hamiltonian does not necessarily reproduce the magnitude of the experimental data; (ii) a wave function (e.g. Ψ_A) which does not diagonalize the Hamiltonian is able to do it. Satisfying Kato's cusp conditions is a mathematical requirement of the Schrödinger equation, while making the Hamiltonian diagonal is only one way of removing the Coulomb singularities. Both these properties, however, are not enough to fully deal with the complete Hamiltonian because of the presence of the non-diagonal H'_{12} term which mixes the variables r_1, r_2 and r_{12}: the three-body system cannot be reduced to the product of 3 two-body systems!

A brief comment of the influence of orthogonalizing the final state Ψ_f^- to the initial state Ψ_i is necessary. For all the initial state wave functions considered above, the magnitude changes by about $10-15\%$ (as stated in [3]) without affecting the shapes. From this point of view, we may thus consider the 3C model to be satisfactory in these geometries and kinematics.

We have further investigated the comparison experiment-theory with a simulation which takes into account the experimental accuracy on the scattering angle of the scattered electron ($\Delta\theta_s = 0.02^0$). An illustration of the results is shown in the top panel of Fig. 3 for $\theta_a = 319^0$: the variations due to the accuracy on θ_s cannot explain the magnitude discrepancy between theory and experiments in the case of Hylleraas-type functions. It would be interesting to make also a simulation which takes into account the energy resolution of the ejected electrons ($\Delta E_a = \Delta E_b = 2$eV) by considering the convolution product of the theoretical result at $E_a = E_b = 10$ eV with the experimental resolution function. We have instead considered the extreme cases where ΔE_a and ΔE_b are treated as if they were accuracies (the total energy of the two ejected electrons being kept at $E_a + E_b = 20$ eV). This simulation (bottom panel of Fig. 3) further indicates that the discrepancy cannot be resolved in this way.

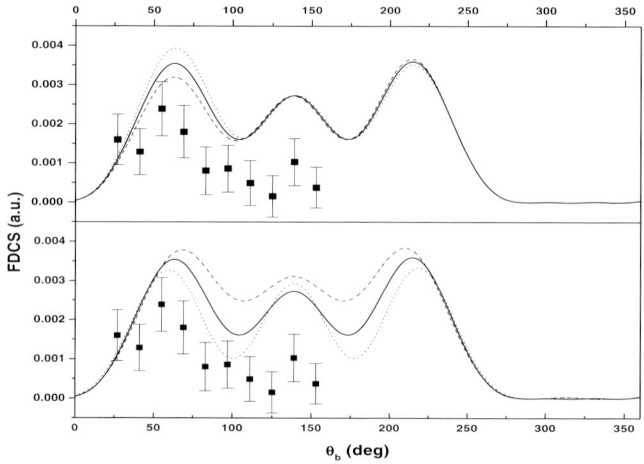

Figure 3. Same as in Fig. 1. All theoretical curves are obtained within the 3C model and with the Bonham and Kohl - number 14 - initial wave functions for the helium ground state. Top panel ($E_a = 10$ eV, $E_b = 10$ eV): $\theta_s = 0.45^0$ (solid line), $\theta_s = 0.43^0$ (dotted line), $\theta_s = 0.47^0$ (dashed li ne). Bottom panel ($\theta_s = 0.45^0$): $E_a = 10$ eV, $E_b = 10$ eV (solid line), $E_a = 8$ eV, $E_b = 12$ eV (dotted line), $E_a = 12$ eV, $E_b = 8$ eV (dashed line).

5. Concluding remarks

Many questions remain open with respect to the interpretation of the (e,3e) measurements [2] on helium at ~ 5.6 keV incident energy. Amongst them we may ask: (i) is the 3C model for the final state sufficient? (ii) why are the CCC results off in magnitude by a factor 3? In this contribution, we have not answered these questions, but learned that:

- within the 3C model, the FBA seems sufficient for the kinematical situation.
- within the 3C model, the absolute value of the calculated cross sections is very sensitive to the used helium bound state wave functions, with variations in magnitude up to a factor 2. The overall good agreement with the absolute experimental data can be equally obtained with Ψ_P or Ψ_A. The study with two other diagonal wave functions (Ψ_0 and Ψ_1) has demonstrated that the conclusions of Jones and Madison [3] are not correct: the fact that the bound state wave function diagonalizes the Hamiltonian *is not* the deciding factor in obtaining such an agreement.
- a simulation which takes into account the experimental accuracy on the scattering angle of the scattered electron and, somehow, the energy resolution of the ejected electrons cannot explain the discrepancy between theory and experiments.

Further theoretical investigations are clearly needed, in particular with respect to the description of the final state. Also, new absolute experimental measurements in similar kinematical and geometrical situations would be welcome.

Acknowledgments

The authors would like to thank A. Lahmam-Bennani for useful comments and for providing experimental data, and the CINES for providing computing time.

References

[1] Berakdar J, Lahmam-Bennani A and Dal Cappello C 2003 *Phys. Rep.* **374** 91

[2] Lahmam-Bennani A *et al.* 1999 *Phys. Rev. A* **59** 3548

[3] Jones J and Madison D H 2003 *Phys. Rev. Lett.* **91** 73201

[4] Brauner M, Briggs J and Klar H 1989 *J. Phys. B: At. Mol. Phys.* **22** 2265

[5] Joulakian J, Dal Cappello C and Brauner M 1992 *J. Phys. B: At. Mol. Phys.* **25** 2863

[6] Hylleraas E A 1928 *Z. Phys.* **48** 469; 1929 **54** 347; 1930 **65** 759

[7] Pluvinage P 1950 *Ann. Physique* **5** 145; *J. Phys. Radium* 1951 **12** 789

[8] Bray I and Stelbovics A T 1992 *Phys. Rev. A* **46** 6995

[9] Kheifets A S *et al.* 1999 *J. Phys. B: At. Mol. Phys.* **32** 5047

[10] Berakdar J 2000 *Phys. Rev. Lett.* **85** 4036

[11] Malcherek A W and Briggs J S 1997 *J. Phys. B: At. Mol. Phys.* **30** 4419

[12] Götz J R, Walter M and Briggs J S 2003 *J. Phys. B: At. Mol. Phys.* **36** L77

[13] Dorn A *et al.* 2003 *Electron and photon impact ionization and related topics 2002* (Bristol and Philadelphia: IOP, L. U. Ancarani ed), p 41

[14] Ancarani L U, Montagnese T and Dal Cappello C 2004 *Phys. Rev. A* **70** 012711

[15] Grin M *et al.* 2000 *J. Phys. B: At. Mol. Phys.* **33** 131

[16] Kheifets A S 2004 *Phys. Rev. A* **69** 32712

[17] Bonham R A and Kohl D A 1966 *J. Chem. Phys.* **45** 2471

[18] Drake G W F 1999 *Physica Scripta* **T83** 83

[19] Bartlett J H, Gibbons J J and Dunn C G 1935 *Phys. Rev.* **47** 679; Bartlett J H 1955 *Phys. Rev.* **98** 1067

[20] Ancarani L U and Popov Y V 2002 *Correlations, polarization, and ionization in atomic systems* (New York: AIP, Madison D H and Schulz M eds) p 115

[21] Kato T 1957 *Comm. Pure Appl. Math.* **10** 151

[22] *Handbook of Mathematical Functions* 1972 (New York: Dover, Abramowitz M and Stegun I A eds)

[23] Le Sech C 1997 *J. Phys. B: At. Mol. Phys.* **30** L47

[24] Popov Y V and Ancarani L U 2000 *Phys. Rev. A* **62** 42702

[25] A further study by these authors [26] has led them to modify their conclusion.

[26] Jones S, Macek J H and Madison D H 2004 *Phys. Rev. A* **70** 012712

Inst. Phys. Conf. Ser. No. 183
Paper presented at Int. Conf. Electron and Photon Impact Ionization and Related Topics, Louvain-la-Neuve, 2004

Modern studies of anisotropy and polarization in Auger-electron emission

N M Kabachnik[1]

Institute of Nuclear Physics, Moscow State University, Moscow 119992, Russia,
Fakultät für Physik, Universität Bielefeld, 33615 Bielefeld, Germany

Abstract. A dramatic improvement in the intensity and quality of the synchrotron radiation beams leads to an explosion of angle and spin resolved investigations of photoinduced Auger processes including coincidence experiments. They aim at a better understanding of the subtle details of the Auger process itself as well as at applications of the angle- and spin-resolved Auger spectroscopy for solving the problems of solid state and molecular physics. A brief overview of recent tendencies in the study of angular distribution and spin polarization of photoinduced Auger electrons is presented.

1. Historical introduction: A tribute to pioneers

The investigations of the angular anisotropy of Auger electron emission were initiated by a pioneering paper of Werner Mehlhorn [1] published in 1968. In his short letter Mehlhorn showed that in the process of ionizing an atomic electron ($j > 1/2$) by particle impact, the ionized atom will be *aligned* with respect to the particle beam direction and the following Auger electrons will have a non-isotropic angular distribution. Shortly after, Cleff and Mehlhorn [2] published first experimental evidence of the anisotropy of Auger electron emission generated by electron impact ionization of Ar(2p). In the following decade the anisotropic emission of Auger electrons were studied using mainly particle (electron, proton) impact ionization (see reviews [3] and references therein). However, already in 1972 Flügge, Mehlhorn and Schmidt [4] noted that the alignment of the ionized atoms can occur also in photoionization even by unpolarized radiation, and therefore the photoinduced Auger electrons should also have non-isotropic angular distribution. Experimentally it was proved almost ten years later by Southworth *et al* [5] who used synchrotron radiation for studying the angular distributions of photo and Auger electrons from Xe(4d). In agreement with the theoretical predictions [4, 6] the anisotropy of Auger electrons produced by photoionization was rather small at all photon energies except in the near-threshold region and near a Cooper minimum. Much larger anisotropy was discovered in resonant Auger process in a pioneering experiment by Carlson *et al* in 1988 [7].

[1] **E-mail: nkabach@mail.ru**

In 1980 Klar has theoretically shown that the Auger electrons should be spin polarized [8] (this was independently found also in [9]). According to Klar, there are two reasons for spin polarization: (i) the alignment of the ion can lead to the *dynamical polarization* due to the spin-orbit interaction, and (ii) the *orientation* of the ion produced by polarized projectile or circularly polarized ionizing radiation causes the Auger electron spin polarization due to the *polarization transfer*. A first measurement of the dynamical spin polarization was made by Hahn *et al* in 1985 [10]. That was a heroic attempt since, as we know now, the dynamical spin polarization generally is very small, of the order of 1% or smaller. Although the measured polarization was marginal, practically within the error bars, this paper stimulated large scale theoretical investigations by two groups: in Münster by Blum, Lohmann and others [11] and by our group in Moscow [12]. Theoretical predictions of a large spin polarization of Auger electrons produced in photoionization by circularly polarized photons [13] were soon confirmed in the first experiments on polarization transfer made by Heinzmann and coworkers for the Auger emission from Rb-layers [14] and from free Ba atoms [15].

All the above mentioned experiments were performed by detecting only one electron. In 1975 McFarlane suggested studying the angular correlations between photo and Auger electrons [16]. He claimed that in this case the anisotropy of Auger electron emission is much larger and is very sensitive to the theoretical parameters. His theoretical derivations were later corrected in our papers [17, 18] but his idea was right. First measurement of the angular correlation between photo and Auger electrons detected in coincidence was successfully realized by Kämmerling and Schmidt (1991) [19] for the photoionization of the Xe $4d_{5/2}$ subshell. Since that time, coincidence measurements of two emitted electrons have become a powerful method of studying electron-electron correlations in atoms, molecules and solids (see below).

During the last decade a dramatic improvement in the intensity and quality of the synchrotron radiation beams has lead to an explosion of angle- and spin-resolved investigations of Auger processes including coincidence experiments. At present, the majority of such investigations are performed with synchrotron radiation beams. Some of the recent achievements in this field will be discussed in this review. References to earlier works may be found in the reviews [20].

2. Theoretical introduction

In this section we shortly remind the known expressions for the angular distribution and spin polarization of Auger electrons mainly with the aim of introducing the necessary notations. The general formalism of angular correlation and polarization measurements in application to the Auger processes was developed on the basis of the density matrix and statistical tensor theory (see, for example, [21]).

In the case of the photoinduced Auger process, the angular distribution of Auger electrons is described by a standard expression [21]

$$\frac{dW}{d\Omega}(\vartheta) = \frac{W_0}{4\pi}\left(1+\beta P_2(\cos\vartheta)\right) \tag{1}$$

where W_0 is the total rate of a considered transition, $P_2(x)$ is the second Legendre polynomial, ϑ is the emission angle relative to the z-axis which is usually chosen along the photon polarization (for linearly polarized light) or along the direction of the photon beam (for circularly polarized light). Within the conventional two-step approach the asymmetry parameter β is presented as a product of two factors $\beta = \alpha_2 \mathscr{A}_{20}$ where $\mathscr{A}_{20}$ is the alignment of the decaying state and α_2 is the intrinsic anisotropy parameter of the Auger decay specific for a particular Auger transition.

The three components of the Auger electron spin polarization can be similarly parameterized. For example, a longitudinal component (along the direction of Auger electron emission) can be presented as [8]

$$P_{\parallel} = \frac{\delta_1 \mathscr{A}_{10} \cos\vartheta}{1 + \alpha_2 \mathscr{A}_{20} P_2(\cos\vartheta)} \quad (2)$$

where $\mathscr{A}_{10}$ is the orientation of the initial state and δ_1 is another intrinsic parameter which characterizes the angular dependence of the longitudinal spin projection. Two other components, perpendicular to the emission direction, lying in the reaction plane and perpendicular to it, can be similarly parameterized with two other intrinsic parameters ξ_1 and ξ_2, respectively [8]. All of the intrinsic parameters can be expressed in terms of bilinear combinations of the Auger amplitudes with coefficients which depend on the angular momenta involved.

In a more elaborate experiment that involves the coincident angular-resolved detection of photo and Auger electrons, more intrinsic parameters are accessible. In general, for a fixed direction of photoelectron emission ($\vec{n}_1$), the angular distribution of the Auger electrons can be presented in the following form [21]

$$I(\vartheta, \varphi; \vec{n}_1) = C\left(1 + \sum_{k=2,4,\ldots} \alpha_k \sum_q \sqrt{\frac{4\pi}{2k+1}} \mathscr{A}_{kq}(\vec{n}_1) Y_{kq}(\vartheta, \varphi)\right) \quad (3)$$

where $Y_{kq}(\vartheta, \varphi)$ are the spherical harmonics. In coincidence experiments, higher-order state multipoles ($2 < k \leq 2J_i$) are formed and therefore higher-order intrinsic parameters depending on Auger amplitudes are accessible from the angular-correlation function. Note that in representation (3), the differential alignment tensors $\mathscr{A}_{kq}(\vec{n}_1)$ depend on the photoionization amplitudes. This fact was used for evaluation of the latter from the experiment (see below).

3. Non-coincidence experiments

3.1. Angular distributions

Since the mid-nineties measurements of the angular anisotropy of Auger and resonant Auger-electron emission have become a routine method of studying the Auger-decay process used practically by all groups involved in such studies. Since the value of the anisotropy parameter α_2 is different for different quantum numbers of the initial and final ionic states, the measurements of the angular distribution can be used as a pure spectroscopic tool for line assignment. This is not so important for diagram Auger transitions, but may be very useful when the Auger spectrum is very complicated as, for example, in the low-energy part of resonant Auger spectra of noble gases [22]. It is also important for high-resolution measurements when theoretical calculations of the line positions are not accurate enough for line identification. As an example, figure 1 shows a part of the resonant Auger spectrum which resulted from $1s^{-1}3p$ photoexcitation of Ne [23]. This high resolution spectrum was measured by the Doppler-free resonant Raman Auger technique which is discussed in detail in [23, 24] (see also contribution to this conference by De Fanis *et al* [25]). Here we only note that the angular distribution of the peaks 1-3 is completely different and a comparison of the measured and calculated anisotropy helped to assign the peaks [23].

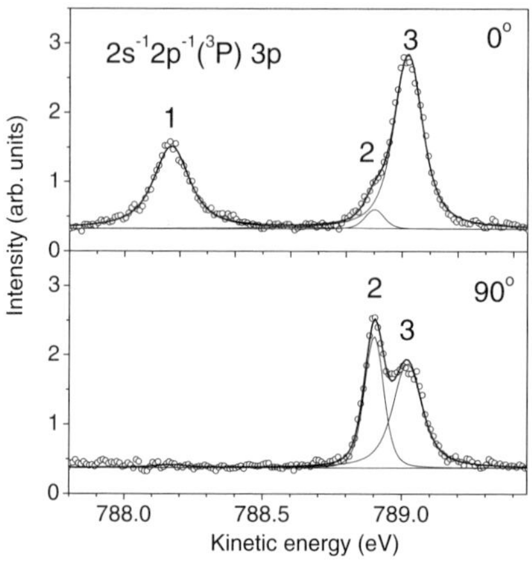

Figure 1. A part of the electron spectra of the resonant Raman Auger transitions from Ne $1s^{-1}3p$ resonance to the Ne^+ $2s2p^5(^3P)3p\,^2S,^2P$ and ^2D states (lines 1,2 and 3, respectively) excited by photons with two directions of linear polarization (see ref. [23] for details).

The main application of the angular anisotropy measurements is, of course, the study of the Auger decay dynamics. Modern high precision measurements of β-parameters helped to understand the limitations of the two-step model and to study the influence of the interference effects [26, 27, 28]. The measurements of the angular anisotropy play an important role in attempts to realize a complete experiment for photoionization or for Auger decay (an experimental determination of complex amplitudes of the process) which will be shortly discussed below. As for the theoretical description of the anisotropy, in general, modern multiconfiguration Dirac-Fock (MCDF) calculations agree quite well with experiments. However, sometimes even for diagram transitions the accurate β-measurements show clear discrepancy with the theory. For example, in the case of Xe $N_{4,5}O_{2,3}O_{2,3}$ transitions to $J=2$ states, even the most elaborate multichannel multiconfiguration Dirac-Fock calculations [29] fail to describe the anisotropy which is known now with high precision [30]. Another example is given by recent measurements of anisotropy for cascade Auger transitions in Ar following 2p $\rightarrow$ 4s photoexcitation made by two groups [31, 32]. The experimental data of the two groups are in agreement, but the measured anisotropy of the second step Auger transition to some of the final states (e.g. $3p^4\,^1D$) clearly disagree with both calculations. Analysis of this case [33] shows that the main obstacle for obtaining better results is poor convergence of computations involving Rydberg orbitals. It would appear that new computational approaches are necessary in order to increase the configuration space by order(s) of magnitude; for example, the approaches similar to those used in quantum chemistry.

3.2. Spin polarization

Spectacular progress in the field of spin- and angle-resolved Auger electron measurements was achieved during the last decade due to a dramatic increase of the intensity of the beams available at the third-generation synchrotron radiation facilities. In addition, almost 100% linearly and circularly polarized photon beams are available nowadays. For this reason difficult spin-

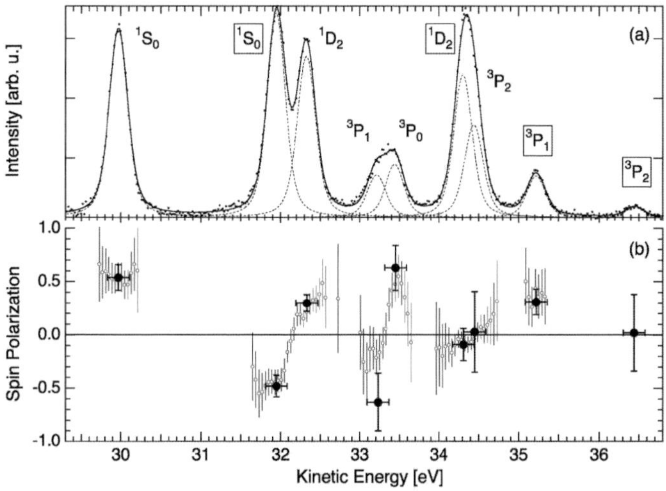

Figure 2. Xe $N_{4,5}O_{2,3}O_{2,3}$ Auger spectrum excited by 93.8 eV photons. (a) Intensity spectrum with fitted Voigt profiles. The framed labels correspond to N_4 initial hole states. (b) Measured spin polarization across the Auger spectrum evaluated for each 40 meV kinetic energy interval. Details see in ref. [35].

polarization measurements for Auger electrons evolved from being an exotic experiment to a useful tool for investigation of Auger electron dynamics. Extensive study of spin polarization of Auger [34, 35, 36] and resonant Auger transitions [37] in noble gas atoms has been undertaken recently. As an example of modern results figure 2 shows the recently measured Xe $N_{4,5}O_{2,3}O_{2,3}$ Auger electron intensity spectrum and corresponding spin polarization obtained using circularly polarized radiation [35]. The spin polarization component along the photon helicity has been measured which originates from the polarization transfer and is proportional to ξ_1 intrinsic parameter[2]. A detailed analysis of the spin-resolved spectra yields the values of ξ_1 for all Auger lines. A comparison of the obtained results with MCDF calculations [38] shows rather good agreement for a majority of the lines. However, as in the case of angular distribution, agreement is poor for the transitions to $J = 2$ final ionic states. New theoretical analysis for these transitions is desirable.

Measurements of Auger electron spin polarization have found very interesting applications in the studies of solid-state magnetization. After pioneering experiments by Landolt and coworkers [39] spin-polarized Auger spectroscopy of magnetically ordered solids has become a powerful probe of local, element-specifi c magnetic properties of surfaces and thin films. In the first experiments unpolarized electron beams were used for generating Auger spectra [39]. In modern experiments on magnetic materials, soft X-ray synchrotron radiation is used (e.g. [40] and references therein). In both cases the main source of spin polarization is the initial magnetization of the solid. On the other hand, spin polarization of Auger electrons from non-magnetic materials induced by circularly polarized light was also studied [14, 41]. In this case the main mechanism of spin polarization is the polarization transfer.

[2] In [35] a notation $\gamma_1/2 - \beta_1$ is used instead of ξ_1 for the intrinsic parameter.

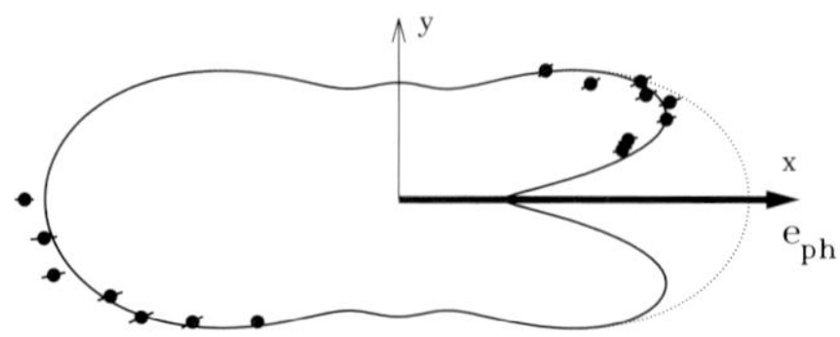

Figure 3. Experimental (dots) and theoretical (curves) angular correlations between photo and Auger electrons in Xe $4d_{5/2}$ photoionization with subsequent $N_5O_{2,3}O_{2,3}\,{}^1S_0$ Auger decay [47] (see text). The direction of the photoelectron emission is along the major axis of light polarization (x-direction).

3.3. Complete experiment for Auger decay

The feasibility of accurate angular distribution and spin-polarization measurements has increased an interest in the realization of the so-called complete experiment for Auger decay, i.e. determination of the Auger-decay amplitudes directly from the experimental data. The concept of the complete experiment for Auger processes was discussed more then 10 years ago [42], however, only recently its realization has become a practical task. Several ways have been explored (see discussion and references in recent review [43]). However, seemingly the most direct way via measurements of the angular anisotropy and the spin polarization parameters has lead to an unexpected result [36]. It was proved that the intrinsic parameters $\alpha_2, \delta_1, \xi_1$ and ξ_2 are not independent and that by measuring only the parameters of the Auger electrons it is not possible, in principle, to realize the complete experiment. Explicit equations relating the Auger intrinsic parameters have been found for many particular cases. However, a general theoretical treatment of this problem is still missing. Now it is clear that in order to realize a complete experiment, information about the polarization state of the residual ion is necessary. Additional information about this subject may be found in the report by Grum-Grzhimailo and Meyer [44] to this conference.

4. Coincidence experiments

4.1. Photoelectron - Auger electron coincidence

Already the first experiment on the photoelectron – Auger electron angular correlation study [19] has demonstrated how informative such measurements are for investigation of the photoionization and Auger processes. That particular experiment on the Xe $4d_{5/2}$ subshell photoionization with the subsequent $N_5O_{2,3}O_{2,3}\,{}^1S_0$ Auger decay was used for determination of the complex photoionization amplitudes. The photoelectron–Auger electron coincidence method was later used by several groups for studying subtle details of the resonant photo double-ionization of atoms such as interference effects, post-collision interaction (PCI) and effects of indistinguishability of emitted electrons at equal energies [45] (see also a contribution to this conference by Rioual *et al* [46]). As an example in figure 3 recent experimental results [47] for angle dependent transition probability in sequential (resonance) photo-double-ionization is shown for the above mentioned process in xenon. The two electrons have been detected in coincidence in a plane perpendicular to the photon beam. The energies of the photoelectron

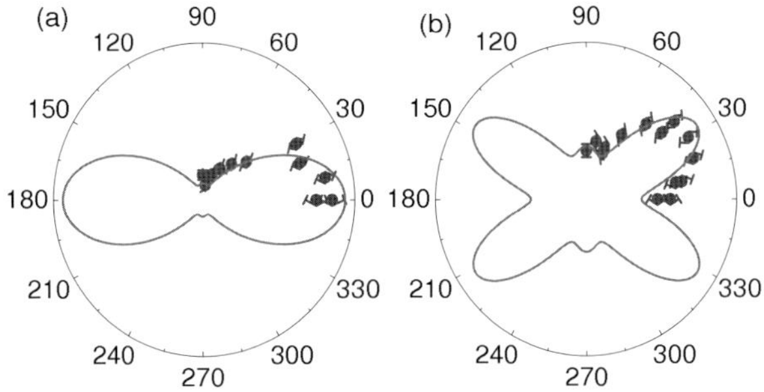

Figure 4. Angular correlation between the resonant Auger emission and the second-step Auger emission for the Kr transitions (a) $3d_{5/2}^{-1}5p$ $(J_0 = 1) \rightarrow 4s^1 4p^5 (^1P)\, 5p \rightarrow 4s^2 4p^4\ ^1D_2$; (b) $3d_{5/2}^{-1}5p$ $(J_0 = 1) \rightarrow 4s^1 4p^5 (^1P)\, 5p \rightarrow 4s^2 4p^4\ ^3P_2$. Coincidence rates measured in a plane perpendicular to the photon beam are plotted as a function of the angle of the direction of first step Auger electron relative to the polarization vector whereas the second step Auger electron is detected in the direction at 270°. The solid curves represent the results of the MCDF calculation. (For details see [54].)

and the Auger electron are 16.97 eV and 29.97 eV, respectively. For comparison the theoretical predictions from the quantum-mechanical model of PCI [48] (solid line) and the shape without PCI [18] (dotted line) have been included. One sees a clear effect of PCI which leads to suppression of the probability of two electron emission at small relative angles.

Very interesting extensions of the angle-resolved photoelectron – Auger electron coincidence spectroscopy have recently been developed for the study of surfaces. The Auger-photoelectron coincidence spectroscopy (APECS) integrated over angles has already been used for rather a long time [49] as a very discriminative site-specific method of solid-state analysis. The first angle-resolved APECS measurements [50] opened new perspectives for the method. The angular distribution of Ge $L_3M_{4,5}M_{4,5}$ Auger electrons in coincidence with Ge $2p_{3/2}$ core photoelectron has been measured. These first measurements demonstrated the increased surface specificity of the coincidence technique in comparison with single Auger or photoelectron measurements. Besides it was shown that the angular correlation pattern is not only determined by the electron diffraction but is also sensitive to the alignment of the core state (as for gas targets) and therefore may be used for its study.

4.2. Cascades of Auger transitions

When an inner-shell electron is not ionized but photoexcited to an empty Rydberg state, the resonant Auger decay occurs which may be followed by another Auger (or Coster-Kronig) transition so that the cascade of resonant Auger – normal Auger transitions is formed. This process may be called double resonant photo-double-ionization. Both emitted Auger electrons can be detected in coincidence in an angle resolved experiment similar to that described above. Theoretically, the angular correlation between two cascade electrons is described by the same equation (3) where the alignment tensors of the intermediate state depend on the Auger amplitudes for the first step decay [51]. As was demonstrated by recent experiments [52, 53, 54], the angular correlation between two cascade Auger electrons is very sensitive to the parameters of both decays. It is especially true for the second step Auger transitions in ions which are difficult to study by other methods. Coincidence measurements drastically increase the alignment

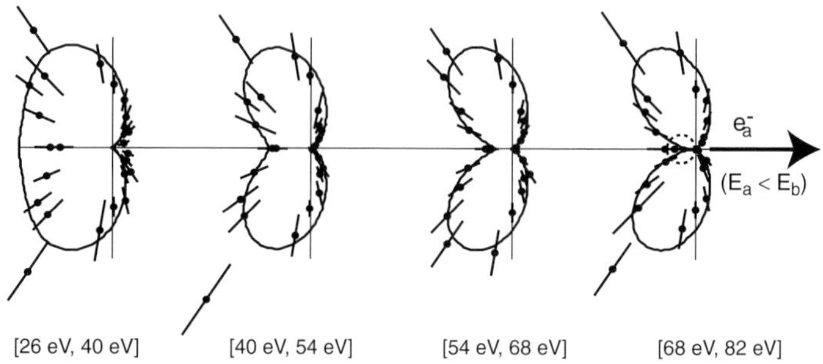

Figure 5. Angular correlation between two Auger electrons in DA decay of Ar $2p^{-1}$ vacancy to all $3p^{-3}$ final ionic states. The slower electron e_a is emitted in the direction indicated by an arrow. The numbers in brackets give the corresponding kinetic energy range of the slower electron (total kinetic energy is 168±8 eV). For details see [56].

of the intermediate state and the anisotropy of the second step decay making it possible to extract the corresponding anisotropy parameter. An example of the correlation patterns for two cascade transitions in Kr are shown in figure 4 [54]. Note that the first stage in both cascades is the same. Thus the observed strong difference in angular correlation patterns is only due to the different intrinsic anisotropy of the second step transitions.

4.3. Double Auger process

The cascade of two subsequent normal Auger transitions is only one of possible vacancy decay channels leading to a triply ionized ion. Another channel is a simultaneous emission of two Auger electrons, a so-called double Auger (DA) process. In the DA decay the transition energy is shared by the two electrons, thus the electron energy spectrum is continual, which makes it difficult to study. Therefore it is not surprising that the DA process is much less studied than the normal Auger decay. Recently the electron-electron coincidence technique most appropriate for DA experiments was applied in investigations of the DA decay of Ne and Ar atoms [55, 56, 57]. Not only the energy distribution but also the angular correlation patterns have been studied. As an example in figure 5 the angular distributions for Ar $L-MMM$ DA transitions are shown for different energy sharing [56]. The correlation patterns show a characteristic two-lobe structure with a minimum at 180° (back-to-back emission). However, even for the energy region close to equal energy sharing (rightmost pattern) there is a contribution of another distribution with a maximum at 180° (shown by dashed line). This contribution increases with decreasing the energy of one of the electrons.

Theoretically the angular correlation in the DA process was first studied by Amusia *et al* [58] for a particular DA transition in Ne. Recently we have made a comprehensive analysis of the possible angular patterns in DA process for various parities (π), spins (S) and orbital angular momenta (L) of the emitted electron pair [59]. A convenient parametrization of the angular distribution $W_{\pi SL}(\theta)$ has been derived which separates kinematical and dynamical correlation factors (θ is a relative angle of two electron emissions). For example, for the $^{2S+1}P^o$ final continuum states, the $W_{^1P^o}(\theta)$ and $W_{^3P^o}(\theta)$ partial angular functions have the form

$$W_{^1P^o}(\theta) = \left|a_{g0}(\theta)\right|^2 (1+\cos\theta) + \left|a_{u0}(\theta)\right|^2 (1-\cos\theta), \quad (4)$$

$$W_{^3P^o}(\theta) = \left|a_{g1}(\theta)\right|^2 (1-\cos\theta) + \left|a_{u1}(\theta)\right|^2 (1+\cos\theta). \quad (5)$$

Here $a_{gS}(\theta)$ and $a_{uS}(\theta)$ are *gerade* and *ungerade* (with respect to the interchange of the two outgoing electrons) amplitudes, respectively. For symmetric energy sharing between the outgoing electrons, the ungerade amplitudes vanish. The dynamical correlation factors, represented by $\left|a_{gS}(\theta)\right|^2$ and $\left|a_{uS}(\theta)\right|^2$, reflect primarily the electron-electron repulsion. In a similarity with the correlation factors for the double photoionization [60], they are approximated by Gaussians as functions of θ with the maximum at $\theta = 180^\circ$. One sees that for equal energy sharing the contribution of the $^1P^o$ state has a typical two-lobe structure with a zero at 180° while that of the $^3P^o$ state has one-lobe structure with a maximum at 180°. An interplay of such contributions explains qualitatively the observed angular patterns.

5. Time-evolution of the Auger process

Finally, I would like to mention a novel experiment where time evolution of the Auger decay was directly investigated [61]. A spectacular development of the physics of short electromagnetic pulses has lead to a possibility of production of single XUV pulses as short as 250 as (1 as = 10^{-18} s). Such ultrashort pulses were produced and analyzed by a collaboration of two groups in Vienna (Krausz) and Bielefeld (Heinzmann) [62]. This pulse duration is already comparable with the time of electron revolution on the Bohr orbital (150 as). It is short enough to make a pump-probe experiment for atomic Auger process feasible. In a recent experiment [61] the life-time of the $d_{5/2}$ vacancy in Kr has been measured in the time domain. The result (7.9 ± 1 fs) is close to the value obtained from the measured spectral width of the Auger line. This experiment was a demonstration of feasibility of such measurements. It opens a new field of investigation, and angle-resolved Auger decay studies which are the subject of this report can contribute considerably to this field.

6. Summary and outlook

The angle- and spin-resolved Auger electron spectroscopy has developed a long way from extremely difficult first experiments on the border of feasibility to a mature almost routine method of investigation of Auger processes. Many interesting problems of the Auger decay in atoms including the problem of a complete experiment have been solved by means of angle- and spin-resolved measurements. I am sure that with the future development of coincidence technique and with preparation of the target atoms (e.g. laser-excited atoms) new interesting physics will appear in the Auger spectroscopy. A new, practically unexplored and very exciting field has appeared recently: the time resolved investigations of the Auger decay which provides complementary information about the process. I believe that in this field we can expect very interesting results. At the same time I think that as happened earlier with the conventional Auger spectroscopy, the angle- and spin-resolved measurements will become a useful method of investigation in molecular, surface and solid-state physics as well as in applied physics. Some of the examples of such applications have been discussed above.

Acknowledgments

I am very grateful to Ulrich Heinzmann, Uwe Becker, Markus Drescher, Alexej Grum-Grzhimailo, Norbert Müller and Kiyoshi Ueda for many stimulating and interesting discussions. I am indebted to Gyorgy Snell, Volker Schmidt and Jens Viefhaus for sending me the figures from their papers and for help. I thank Kay Lofthouse for correcting the manuscript. The hospitality and financial support of Bielefeld University is gratefully acknowledged.

References

[1] Mehlhorn W 1968 *Phys. Lett.* A **26** 166

[2] Cleff B and Mehlhorn W 1971 *Phys. Lett.* A **37** 3

[3] Mehlhorn W 1985 *Atomic Inner-Shell Physics* ed B Crasemann (New York: Plenum Press) p 119; 1990 *X-ray and Inner-Shell Processes* ed T A Carlson *et al* (New York: AIP) v 215 p 465

[4] Flügge S, Mehlhorn W and Schmidt V 1972 *Phys. Rev. Lett.* **29** 7

[5] Southworth S H, Kobrin P H, Truesdale C M, Lindle D, Owaki S and Shirley D A 1981 *Phys. Rev.* A **24** 2257
Southworth S H, Becker U, Truesdale C M, Kobrin P H, Lindle D, Owaki S and Shirley D A 1983 *Phys. Rev.* A **28** 261

[6] Berezhko E G, Kabachnik N M and Rostovsky V S 1978 *J. Phys. B: At. Mol. Phys.* **11** 1749

[7] Carlson T A, Mullins D R, Beall C E, Yates B W, Taylor J W, Lindle D W, Pullen B P and Grimm F A 1988 *Phys. Rev. Lett.* **60** 1382; 1989 *Phys. Rev.* A **39** 1170

[8] Klar H 1980 *J. Phys. B: At. Mol. Phys.* **13** 4741

[9] Kabachnik N M 1981 *J. Phys. B: At. Mol. Phys.* **14** L337

[10] Hahn U, Semke J, Merz H and Kessler J 1985 *J. Phys. B: At. Mol. Phys.* **18** L417

[11] Blum K, Lohmann B and Taute E 1986 *J. Phys. B: At. Mol. Phys.* **19** 3815
Lohmann B 1990 *J. Phys. B: At. Mol. Phys.* **23** 3147

[12] Kabachnik N M and Sazhina I P 1984 *J. Phys. B: At. Mol. Phys.* **17** 1335; 1986 *Opt. Spektrosk. (USSR)* **60** 1106; 1988 *J. Phys. B: At. Mol. Opt. Phys.* **21** 267

[13] Kabachnik N M and Lee O V 1989 *J. Phys. B: At. Mol. Opt. Phys.* **22** 2705

[14] Stoppmanns P, Schmiedeskamp B, Vogt B, Müller N and Heinzmann U 1992 *Phys. Scr.* T **41** 190

[15] Kuntze R, Salzmann M, Böwering N and Heinzmann U 1993 *Phys. Rev. Lett.* **70** 3716

[16] McFarlane S C 1975 *J. Phys. B: At. Mol. Phys.* **8** 895

[17] Berezhko E G and Kabachnik N M 1979 *J. Phys. B: At. Mol. Phys.* **12** 2993

[18] Kabachnik N M 1992 *J. Phys. B: At. Mol. Opt. Phys.* **25** L389

[19] Kämmerling B and Schmidt V 1991 *Phys. Rev. Lett.* **67** 1848

[20] Aksela H, Aksela S and Kabachnik N 1996 *VUV and Soft X-Ray Photoionization* ed U Becker and D A Shirley (New York: Plenum Press) Chap. 11
Mehlhorn W 1998 *J. Electron Spectrosc. Relat. Phenom.* **93** 1; 2000 *X-Ray and Inner-Shell Processes* ed R W Dunford *et al* (New York: AIP) v 506, p 33
Armen G B, Aksela H, Åberg T and Aksela S 2000 *J. Phys. B: At. Mol. Opt. Phys.* **33** R49
Kabachnik N M and Ueda K 2004 *Phys. Scr.* T **110** 73

[21] Balashov V V, Grum-Grzhimailo A N and Kabachnik N M 2000 *Polarization and Correlation Phenomena in Atomic Collisions. A Practical Theory Course* (New York: Kluwer Academic/Plenum)

[22] Becker U and Wehlitz R 1994 *J. Electron Spectrosc. Relat. Phenom.* **67** 341

[23] Ueda K *et al* 2003 *Phys. Rev. Lett.* **90** 153005

[24] De Fanis A, Tamenori Y, Kitajima M, Tanaka H and Ueda K 2004 *J. Electron Spectrosc. Relat. Phenom.* **137-140** 271

[25] De Fanis A ,Tamenori Y, Kitajima M, Tanaka H and Ueda K 2004 contribution to this volume

[26] Ueda K, Shimizu Y, Kabachnik N M, Sazhina I P, Wehlitz R, Becker U, Kitajima M and Tanaka H 1999 *J. Phys. B: At. Mol. Opt. Phys.* **32** L291

[27] Kitajima M *et al* 2001 *J. Phys. B: At. Mol. Opt. Phys.* **34** 3829

[28] Heinäsmäki S 2001 *J. Electron Spectrosc. Relat. Phenom.* **114-116** 75.

[29] Tulkki J, Kabachnik N M and Aksela H 1993 *Phys. Rev.* A **48** 1277

[30] Snell G, Kukk E, Langer B and Berrah N 2000 *Phys. Rev.* A **61** 042709

[31] Ueda K, Shimizu Y, Chiba H, Kitajima M, Tanaka H, Fritzsche S and Kabachnik N M 2001 *J. Phys. B: At. Mol. Opt. Phys.* **34** 107

[32] Huttula S-M, Heinäsmäki S, Aksela H, Tulkki J, Kivimäki A, Jurvansuu M and Aksela S 2001 *Phys. Rev.* A **63** 032703

[33] Huttula S-M, Heinäsmäki S, Aksela H, Jurvansuu M and Aksela S 2001 *J. Phys. B: At. Mol. Opt. Phys.* **34** 2325

[34] Snell G, Drescher M, Müller N, Heinzmann U, Hergenhahn U, Viefhaus J, Heiser F, Becker U and Brookes N B 1996 *Phys. Rev. Lett.* **76** 3923
Schmidtke B, Khalil T, Drescher M, Müller N, Kabachnik N M and Heinzmann U 2001 *J. Phys. B: At. Mol. Opt. Phys.* **34** 4293
Snell G, Langer B, Young A T and Berrah N 2002 *Phys. Rev.* A **66** 022701

[35] Snell G, Drescher M, Müller N, Heinzmann U, Hergenhahn U and Becker U 1999 *J. Phys. B: At. Mol. Opt. Phys.* **32** 2361

[36] Schmidtke B, Khalil T, Drescher M, Müller N, Kabachnik N M and Heinzmann U 2000 *J. Phys. B: At. Mol. Opt. Phys.* **33** 5225

[37] Hergenhahn U, Snell G, Drescher M, Schmidtke B, Müller N, Heinzmann U, Wiedenhöft M and Becker U 1999 *Phys. Rev. Lett.* **82** 5020
Drescher M, Khalil T, Müller N, Fritzsche S, Kabachnik N M and Heinzmann U 2003 *J. Phys. B: At. Mol. Opt. Phys.* **36** 3337

[38] Lohmann U, Hergenhahn U and Kabachnik N M 1993 *J. Phys. B: At. Mol. Opt. Phys.* **26** 3327

[39] Landolt M and Mauri D 1982 *Phys. Rev. Lett.* **49** 1783
Allenspach R, Mauri D, Taborelli M and Landolt M 1987 *Phys. Rev.* B **35** 4801

[40] Kucherenko Yu, Sinković B, Shekel E, Rennert P and Hulbert S 2000 *Phys. Rev.* B **62** 5733

[41] David R, Scheuer M, Stoppmanns P, Yu S-W, Müller N and Heinzmann U 1998 *J. Electron Spectrosc. Relat. Phenom.* **93** 153

[42] Kabachnik N M and Sazhina I P 1990 *J. Phys. B: At. Mol. Opt. Phys.* **23** L353

[43] Kabachnik N M 2004 *J. Electron Spectrosc. Relat. Phenom.* **137-140** 305

[44] Grum-Grzhimailo A N and Meyer M 2004 contribution to this volume

[45] Viefhaus J, Snell G, Hentges R, Wiedenhöft M, Heiser F, Geßner O and U. Becker 1998 *Phys. Rev. Lett.* **80** 1618
Scherer N, Lörch H, Kerkau T and Schmidt V 1999 *Phys. Rev. Lett.* **82** 4615
Rioual S, Rouvellou B, Avaldi L, Battera G, Camilloni R, Stefani G and Turri G 2001 *Phys. Rev. Lett.* **86** 1470

[46] Rioual S, Rouvellou B, Salou M, Huetz A and Avaldi L 2004 contribution to this volume

[47] Scherer N, Lörch H, Kerkau T and Schmidt V 2004 *J. Phys. B: At. Mol. Opt. Phys.* **37** L121

[48] Sheinerman S A and Schmidt V 1997 *J. Phys. B: At. Mol. Opt. Phys.* **30** 1677

[49] Haak H W, Sawadzky G A and Thomas T D 1978 *Phys. Rev. Lett.* **41** 1825

[50] Gotter R, Ruocco A, Butterfield M T, Iacobucci S, Stefani G and Bartynski R A 2003 *Phys. Rev.* B **67** 033303

[51] Kabachnik N M, Sazhina I P and Ueda K 1999 *J. Phys. B: At. Mol. Opt. Phys.* **32** 1769

[52] Wehlitz R, Pibida L S, Levin J C and Sellin I A 1999 *Phys. Rev.* A **59** 421

[53] Turri G, Battera G, Avaldi L, Camilloni R, Coreno M, Ruocco A, Colle R, Simonucci S and Stephani G 2001 *J. Electron Spectrosc. Relat. Phenom.* **114-116** 199

[54] Ueda K *et al* 2003 *J. Phys. B: At. Mol. Opt. Phys.* **36** 319

[55] Viefhaus J 2003 *X-Ray and Inner-Shell Processes* ed A Bianconi *et al* (New York: AIP) p 307

[56] Viefhaus J, Cvejanović S, Langer B, Lischke T, Prümper G., Rolles D, Golovin A V, Grum-Grzhimailo A N, Kabachnik N M and Becker U 2004 *Phys. Rev. Lett.* **92**, 083001

[57] Viefhaus J, Grum-Grzhimailo A N, Kabachnik N M and Becker U 2004 *J. Electron Spectrosc. Relat. Phenom.* in press

[58] Amusia M Ya, Lee I S and Kilin V A 1992 *Phys. Rev.* A **45** 4576

[59] Grum-Grzhimailo A N and Kabachnik N M 2004 *J. Phys. B: At. Mol. Opt. Phys.* **37** 1879

[60] Briggs J S and Schmidt V 2000 *J. Phys. B: At. Mol. Opt. Phys.* **33** R1

[61] Drescher M, Hentschel M, Kienberger R, Uiberacker M, Yakovlev V, Scrinzi A, Westerwalbesloh Th, Kleineberg U, Heinzmann U and Krausz F 2002 *Nature* **419** 803
Drescher M, Hentschel M, Kienberger R, Uiberacker M, Westerwalbesloh Th, Kleineberg U, Heinzmann U and Krausz F 2004 *J. Electron Spectrosc. Relat. Phenom.* **137-140** 259

[62] Hentschel M, Kienberger R, Spielmann Ch, Reider G A, Milosevic N, Brabec T, Corkum P, Heinzmann U, Drescher M and Krausz F 2001 *Nature* **414** 509
Kienberger R, Goulielmakis E, Uiberacker M, Baltuska A, Yakovlev V, Bammer F, Scrinzi A, Westerwalbesloh Th, Kleineberg U, Heinzmann U, Drescher M and Krausz F 2004 *Nature* **427** 817

Inst. Phys. Conf. Ser. No. 183
Paper presented at Int. Conf. Electron and Photon Impact Ionization and Related Topics, Louvain-la-Neuve, 2004

Coherence effects in Auger-photoelectrons coincidence experiments

S Rioual[1], B Rouvellou[1], M Salou[1], A Huetz[2], and L Avaldi[3]
[1] LCEA - Université de Bretagne Occidentale; 6 av. V. le Gorgeu; 29285 Brest Cédex France
[2]LIXAM UMR 8624, Bat. 350, Centre d'Orsay, 91405 Orsay Cédex, France
[3]CNR –IMIP, Area della Ricerca di Roma 1, CP10 , 00016 Monterotondo Scalo, Italy

Abstract. The indirect double photoionization of Kr via the $Kr^{+}(3d^{-1}{}_{3/2,5/2})$ states to the $Kr^{2+}(4s4p^{5}\,{}^{1}P^{o})$ final state has been studied by Auger electron-photoelectron angular resolved coincidence experiments. By a suitable tuning of the incident photon energy we succeeded in observing interference effects due to i) the indistinguishability of both electrons in the final state and ii) the coherence between the two non-degenerate spin orbit states. These observations have been described quantitatively by a model which accounts for the two pathways to the final state as well as for the indistinguishability of the two ejected electrons.

1. Introduction

Investigation of coherence in atomic collisions and in photoionization provides fundamental information for the understanding of the dynamics of few charged particles. Due to the advances of synchrotron radiation sources and of electron spectrometers, the experimental investigation of the indirect double photoionization of rare gases has made possible recently the observation of new coherence effects due to wave nature of electrons. The indirect double photoionization is characterized by the formation of an excited state of the singly charged ion that may decay into the double continuum:

$$h\nu + A \rightarrow A_{1}^{+} + e_{ph} \rightarrow A^{2+} + e_{ph} + e_{Au} \qquad (1)$$

Usually, the photoelectron ejected in the primary inner shell ionization and the Auger electron ejected in the decay of the inner hole are distinguished by their kinetic energies and the process is interpreted as a two step process, in which the emission of two electrons is treated separately. However interference effects due to electron exchange have been predicted in the particular case when both electrons have approximately the same energy [1]. In such a case (noted hereafter as interference case 1), the two step approximation has to be replaced by a one step approach, where the two electrons are treated on equal footing. Thus the coherent summation of two amplitudes where the two electrons exchange their role of photoelectron-Auger electron, gives rise to interference effects which are the signature of their indistinguishability. In the last years, electron-

electron coincidence experiments performed with an energy resolution comparable to the natural width of the intermediate state have indeed allowed to observe such effects in the angle- and energy-dependence of the cross section [2, 3, 4].

Another kind of interference (noted hereafter as interference case 2), may appear by considering that indirect double photoionization may also proceed through many different intermediate excited states to the same final state. Such coherence effects between different excited states have been observed in electron impact experiments where the excitation of neutral states embedded in the ionization continuum and their autoionization have been studied. The first experiments were performed near threshold [5], where the postcollisional interaction, which broadens and shifts the energy distribution of scattered and ejected electrons, makes to overlap the electron distributions from two different states. Then, Van den Brink et al. [6, 7] developed a method to observe the interference effects due to coherence between non-degenerate states far from threshold. In these experiments on helium, the incident energy was chosen such that the energy of the scattered electron from one autoionizing state was equal to the energy of the ejected electron from a different autoinizing state. Photoionization experiments offer a few advantages in comparison to the electron impact ones. In photoionization the contribution of the direct double ionization is negligible and the number of angular momenta of the photoelectron and the Auger electron are limited by selection rules. This makes the analysis of the experiment and the extraction of the relevant physical parameters more straightforward. Pioneering investigations in photoionization were performed by de Gouw et al. [8, 9]. In those non coincident experiments only one electron was detected. Thus the measured cross section was integrated over all directions of the undetected electron and, therefore, such experiments cannot be considered as kinematically complete. In the present study, we will prove that electron-electron coincidence experiments allow to provide unambiguously the phase difference between these two states. The studied process is then described by:

$$h\nu + A \rightarrow A_1^+ + e_{ph}^{(1)} \rightarrow A^{2+} + e_{ph}^{(1)} + e_{Au}^{(1)} \qquad (2\text{-}1)$$

$$h\nu + A \rightarrow A_2^+ + e_{ph}^{(2)} \rightarrow A^{2+} + e_{ph}^{(2)} + e_{Au}^{(2)} \qquad (2\text{-}2)$$

The same final state of the doubly charged ion can be reached via two different intermediate states A_1^+ and A_2^+, and the photon energy is chosen in a way that the energy of the photoelectron $e_{ph}^{(1)}$ becomes equal to the energy of the Auger electron $e_{Au}^{(2)}$, implying the same condition for the other pair of electrons $e_{ph}^{(2)}$, $e_{Au}^{(1)}$. In this way the final states of the two processes become indistinguishable and interferences should occur.

As stated by Viefhaus et al. [3], to observe interference effects in the coincidence angular and energy distributions, the challenge is to achieve an energy resolution comparable or even smaller than the natural linewidth of the intermediate state. Therefore, we have chosen to work with relatively low photon energies with the spin-orbit $3d^{-1}{}_{3/2,5/2}$ doublet acting as the intermediate states

$$\begin{aligned} h\nu + Kr &\rightarrow Kr^{+}3d^{-1}(^2D_{3/2}) + e^1_{ph}(l=1,3) \\ &\rightarrow Kr^{2+}4s4p^5(^1P^o) + e^1_{Au}(l=1,3; E^1_{Au}=39.08eV) \end{aligned} \qquad (3\text{-}1)$$

$$\begin{aligned} h\nu + Kr &\rightarrow Kr^{+}3d^{-1}(^{2}D_{5/2}) + e^{2}_{ph}(l=1,3) \\ &\rightarrow Kr^{2+}4s4p^{5}(^{1}P^{o}) + e^{2}_{Au}(l=1,3; E^{2}_{Au}=37.84eV) \end{aligned} \tag{3-2}$$

and to select their decay to the Kr^{2+} ($4s^14p^5$) [$^1P^{o}$] final state with the ejection of the $M_{4,5}N_1N_{23}$ Auger electrons. Both the photoelectron and Auger electron have the same angular momentum. Thus by tuning the energy of the photon one can produce a final state in which the electrons are completely indistinguishable as far as their kinetic energy and angular momentum are concerned.

In order to perform a complete investigation of such effects, we have measured the coincidence angular distributions corresponding to the following cases :

i) the angular distribution of a photoelectron emitted from process (3-1) (the corresponding Auger electron being detected at a fixed direction);

ii) the angular distribution of an Auger electron ejected from the process (3-2) (the corresponding photoelectron being detected at a fixed direction);

iii) the angular distribution of the mixed Auger electron (3-2) / Photoelectrons (3-2). Interference case 1.

iv) the angular distribution of the mixed Auger electron (3-2) / Photoelectrons (3-1). Interference case 2.

2. Experimental set-up

The experiments have been performed at the Gas Phase Photoemission beam–line of the Elettra storage ring using the multicoincidence end-station [10] which houses two independently rotatable arrays of hemispherical electrostatic spectrometers located in the plane perpendicular to the direction of propagation of the incident radiation. In the present measurements, one electron, labeled "*A*", was detected in the direction of polarization of the incident radiation (ϑ_A =0°). The angular distribution of the "*B*" electron was obtained by successive rotations of the frame containing seven analyzers. The energy resolution and the angular acceptance in the dispersion plane of the spectrometers were $\Delta E/E$=0.003 and ±3°, respectively. Thus, for energies of about 40 eV, ΔE was about 120 meV comparable to the linewidths of the Kr^{+} $3d^{-1}$ lines ($\Gamma_{3/2,5/2}$=98±12 meV [11]). The angular efficiency of the different analyzers was determined via the measurement of the angular distribution of Kr(3d) at about 132 eV photon energy [12,13].

Figure 1 shows the non-coincidence electron spectra measured at three different photon energies. At hv=132.15 eV, the two main peaks in the spectrum correspond to the photoionization of the 3d states, while the other two are assigned to the $M_{4,5}N_1N_{23}$ Auger peaks [14]. By tuning the photon energy to 131.63 eV, the energy of the photoelectron emitted from process (3-2) is equal to the energy of the Auger electron from the same process. Then, we are in the first case of interference effects due to electron exchange. By tuning the photon energy to 132.88 eV, the $3d_{3/2}^{-1}$ photoline is made to overlap the $M_5N_1N_{23}$ Auger peak at 37.84 eV kinetic energy, and conversely the $3d_{5/2}^{-1}$ photoline now superimposed to the $M_4N_1N_{23}$ Auger peak. This is the condition where the coherence between the two non-degenerate intermediate states should occur. However, the analysis of these spectra does not exhibit any interference effect. This is consistent with the pioneering non coincident measurements of de Gouw *et al.* [9] which display such effects only from the observation of slight variations of the Auger peak position and not from the intensity of the Auger/photoelectron peaks. Note that this result is in disagreement with the theoretical prediction [15] which predicts interference effects, even in non coincidence experiments, when the two electrons in the final state have the same angular momenta. This apparent discrepancy is however fully explained by the value of the Auger/photoelectron branching ratio assumed to be 1 in the theory, but shown to be about 0.1 in our experiments. Therefore to observe a “detectable” interference effects in non coincidence experiments one has firstly to choose a case where the two electrons have the same angular momentum in the final state and secondly the considered decay channel has a high probability.

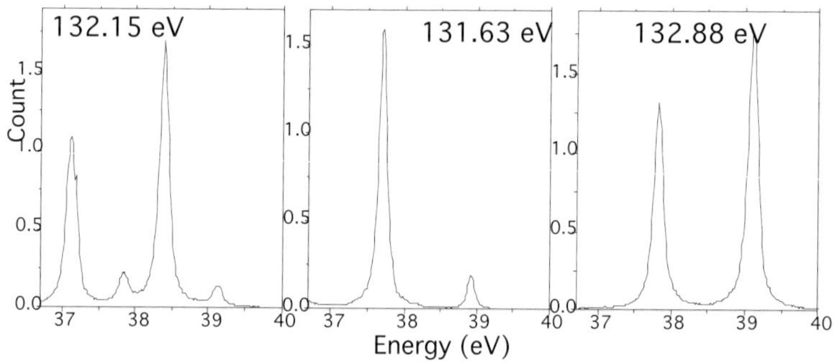

Figure 1. Kr Auger electron / photoelectron spectra measured at 132.15 , 131.63 and 132.88 eV photon energy, respectively.

3. Theoretical approach

According to the formulation of Vegh and Macek [1] and Vegh [15], four different amplitudes contribute to the calculation of the triple differential cross section (TDCS) of processes (3-1) and (3-2). Thus, describing the intermediate and final states of the ion within the LSJ coupling scheme by the respective sets of quantum numbers $L_r S_r J_r$ and $L_f S_f J_f$, the TDCS can be written as follows:

$$TDCS(k_A,k_B) = \sum_{M_f} \left| \frac{A^1(k_A,k_B)}{E_A - E^1_{Au} + \frac{i\Gamma^1}{2}} + \frac{A^1(k_B,k_A)}{E_B - E^1_{Au} + \frac{i\Gamma^1}{2}} + \frac{A^2(k_A,k_B)}{E_A - E^2_{Au} + \frac{i\Gamma^2}{2}} + \frac{A^2(k_B,k_A)}{E_B - E^2_{Au} + \frac{i\Gamma^2}{2}} \right| \quad (4)$$

where the amplitudes A_i (i=1 or 2) for the two processes are given by the equation (5).

$$A_{1,2}(k_A,\sigma_A,k_B,\sigma_B) = \sum_{M_r} \left\langle k_A\sigma_A Kr^{2+}(L_f S_f J_f)\middle| H \middle| Kr^{+}(L_r S_r J_r)\right\rangle \times \left\langle k_B\sigma_B Kr^{+}(L_r S_r J_r)\middle| D\middle| Kr\right\rangle \quad (5)$$

The momenta $\mathbf{k_A}$, $\mathbf{k_B}$,the kinetic energies E_A, E_B and spin projections σ_A, σ_B characterize the two detected electrons as defined above. $E_{Au,i}$ and Γ_i (i=1,2) characterize the intermediate states of the two processes (3-1) and (3-2). D is the dipole operator and H, the Hamiltonian of Kr^+, determines the Auger decay. The first and second terms in (4) represented with A_1 refer to process (3-1) with $J_r = 3/2$ and account for the exchange of the photo- and Auger-electrons, with a minus sign due to antisymmetrization. The third and fourth terms in (4) play the same role, but for the second process (3-2) where J_r=5/2.

The amplitude A_i can be further expanded [16]:

$$\mathrm{A_i}(k_A,\sigma_A,k_B,\sigma_B) = \sum_{L\,S\,J\,M_L\,M_S\,M_J} \sqrt{2L+1}(2S+1)(2J+1)(-1)^J \begin{pmatrix} 1/2 & 1/2 & S \\ \sigma_B & \sigma_A & -M_S \end{pmatrix}$$
$$\times \begin{pmatrix} S & L & J \\ M_S & M_L & -M_J \end{pmatrix} \begin{pmatrix} J & 1 & L_f \\ M_J & 0 & M_f \end{pmatrix} \begin{Bmatrix} 1/2 & 1/2 & S \\ L_r & L_r & J_r \end{Bmatrix} \quad (6)$$
$$\times \sum_{l_A l_B} D_i^{l_B}(k_B) H_i^{l_A}(k_A) \begin{Bmatrix} 1 & L_f & J \\ L_r & L_r & S \\ l_B & l_A & L \end{Bmatrix} Y_{L M_L}^{l_A l_B}(k_A,k_B)$$

H_i^l and D_i^l are the reduced matrix elements for the electrostatic and the dipolar interactions respectively. The parentheses designate the usual 3j, 6j and 9j symbols. Two partial waves (p and f) have to be considered for the photoelectron, and two other ones (p and f) for the Auger electron. The p partial wave can be neglected in the calculations of the Auger decay since the $f_{5/2}$ and $f_{7/2}$ partial waves dominate [17]. Then, in the calculations, we have considered the modulus $\rho$of the ratio of the dipolar matrix elements D^p and D^f and their relative phase Δ. Using the formalism of Cooper and Zare [18] and Berezhko *et al.* [19], we have obtained ρ=0.34 and Δ=0° from earlier measurements of the photoelectron asymmetry parameter β [13] and of the alignment parameter A_{20} [20]. Finally, the $\Gamma_{1,2}$ values reported in the literature [11] and the measured branching ratio $R=|A_2/A_1|^2=1.56\pm0.1$ (Fig.1) for the two spin orbit components have been used. The phase difference between the amplitudes of the two processes which contains the coherence between the two intermediate states has been included in the calculations. It is noteworthy that in the calculation of A_1 and A_2, the 6j Wigner coefficient changes its sign from J_r=3/2 to 5/2. Therefore, the phase can be written as $\phi+\pi$ where ϕ arises from the reduced matrix elements D_i and, π is purely due to angular momentum algebra.

4. Results

Four different angular coincidence distributions have been measured: two at hν=132.15 eV (Fig. 2(a) and 2(b)) where processes (3-1) and (3-2) can be distinguished, one at 131.63 eV (Fig. 2(c)) where the kinetic energies of the two electrons are equal, E_A=E_B, and equal to the energy of the $M_5N_1N_{23}$ Auger electron (37.84 eV) (interference case 1) and one at hv=132.88 eV (Fig. 2(d)) where the processes (2-1) and (2-2) are fully mixed (interference case 2). In all experiments the electron labeled A was detected along the direction of polarization of the incident radiation. All the experimental angular distributions are on the same relative scale.

In Fig. 2a and 2b, the processes (3-1) and (3-2) have been studied separately respectively. In Fig. 2a, the Auger electron emitted from the process (3-1) is detected along the polarization axis while in Fig. 2b, the photoelectron of the process (3-2) is detected in the same direction. Despite the fact that all the four terms, that compose the amplitude, always contribute to determine the TDCS, only small interference effects occur, as shown by the comparison of the two calculations that sum coherently (full line) and incoherently (dotted line) the four terms in equation (5). Our simple calculations give an overall good representation of the data and are able to depict the observed angular pattern. The disagreement observed at 240° and 120° in Fig. 2b, in which the angular distribution of the $M_5N_1N_{23}$ Auger electron is measured in coincidence with a photoelectron detected along the direction of polarization of the radiation, might be due to the neglecting of the p wave for the Auger electron.

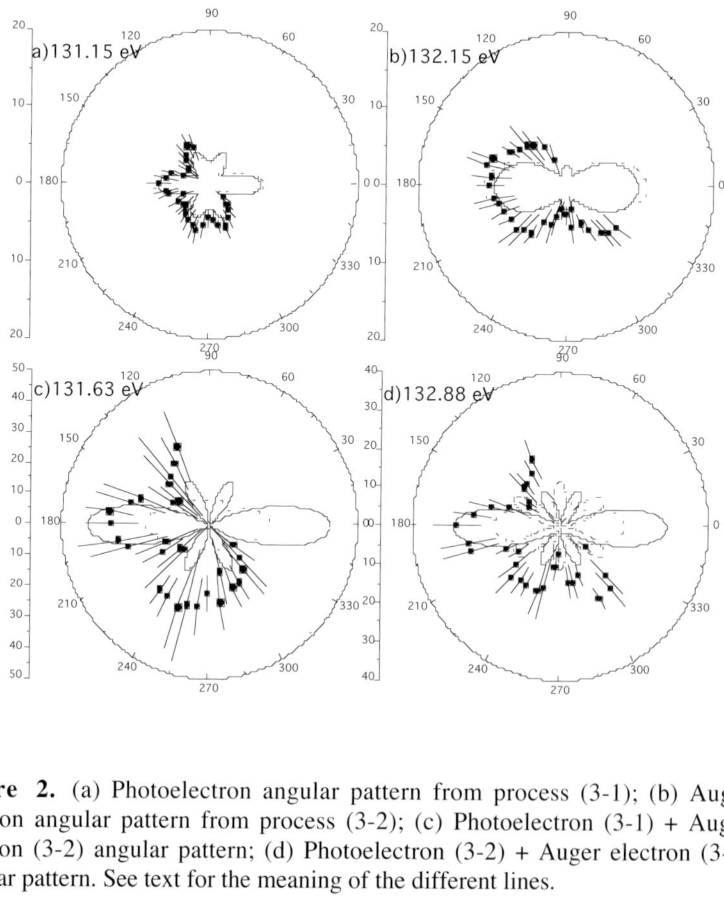

Figure 2. (a) Photoelectron angular pattern from process (3-1); (b) Auger electron angular pattern from process (3-2); (c) Photoelectron (3-1) + Auger electron (3-2) angular pattern; (d) Photoelectron (3-2) + Auger electron (3-2) angular pattern. See text for the meaning of the different lines.

The angular distribution measured at hν=131.63 eV when two electrons of 37.84 eV are detected in coincidence is shown in Fig. 2c. This corresponds to the study of the process (3-2) in a condition where the photoelectron and Auger electron are indistinguishable. Then exchange between the two electrons is responsible for the observed interference effects. The dotted and full lines represent the incoherent and coherent sum of the amplitudes respectively. At 180°, the intensity given by the full line is two times the intensity given by the incoherent sum of the amplitudes defining the process (3-2). Thus a constructive interference is observed. The sign of the interference effect depends on the total spin and parity of the electron pair in the final state. In our case the final ion state is $^1P^o$ and within the LS coupling scheme, three symmetries ($^1S^e$, $^1P^e$, $^1D^e$) of the electron pair exist. The $^1P^e$ symmetry does not contribute at all to the TDCS for two electrons with the same energy, while the two others symmetries of favoured type result in a constructive interference. The same effects have been observed previously in Xenon by Selles et al. [16], who also discussed in detail their link to direct double photoionization.

Finally, in figure 2(d) when the energy of the incident photon is set to 132.88 eV, then the kinetic energy of the Auger electrons produced in process (3-1) is equal to the photoelectron energy in process (3-2) and vice-versa. Now, the first and fourth terms in the expression of the TDCS mainly contribute to the observed angular distribution. The other two terms, which accounts for the exchange of the two electrons, give negligible contribution because of the different kinetic energies of the two electrons E_A and E_B. The figure clearly displays that i) a calculation which sum incoherently the contribution of the processes (3-1) and (3-2) (dotted line in Fig 2(d)) does not reproduce the experimental findings, ii) the shape of the calculated TDCS is strongly dependent on the phase difference ϕ between the amplitudes A_1 and A_2. The dash and full lines represent the results of the calculations with a relative phase of 0° and 135°, respectively. The full line gives a maximum at 180° in agreement with the experimental data whereas the dashed line presents a node at this angle. It is worthwhile to note that the value of 135° corresponds to the best fitting to the data and is in agreement with the results derived by de Gouw et al. [9] in their parametric representation of the non coincidence cross section. However, the extraction of the phase in coincidence experiments is definitely more reliable as shown in Fig. 2(d) which demonstrates the strong dependence of our measurements on ϕ.This result is strongly in contrast with the weakness of the interference effects previously observed by de Gouw et al.[9], and proves the unique role of the electron-electron coincidence technique in the investigation of coherence effects. We emphasize that the constructive interference observed at 180° on both figures 2c and 2d are of completely different nature. In Fig. 2d (case 2) they are related with the first and fourth terms in the TDCS expression (4), and occur because two different pathways (3-1 and 3-2) interfere with each other. The phase between them comes partly from angular algebra and partly from radial matrix elements as discussed above. In Fig. 2c (case 1), only one pathway (3-2) contributes through the third and fourth terms in the TDCS, and the phase between them, only depends on the final symmetry of the electron pair.

In summary the indirect double photoionization of Kr has been studied via a series of photoelectron-Auger electron coincidence experiments, which have enabled us to observe i) interference effects due to electron exchange and ii) the interference effects due to the coherence of non degenerate states. These latter experiments have allowed a better and more quantitative description of the observed interference than the one achieved in previous non coincidence experiments by de Gouw et al [9]. It is worthwhile to mention that the new sources of synchrotron radiation and the increased coincidence efficiency of the *ad-hoc* built electron spectrometers allow to perform such basic experiments, comparable in the interference case 2 to the Young's two slit experiment.

Acknowledgements

We acknowledge the assistance of the staff of the gas phase beamline at Elettra. This work was financially supported by the EEC through project N°. ERB FMGE CT95 0022.

References

[1] L. Vegh and J.H. Macek, Phys. Rev. A **50**, 4031 (1994).

[2] S.J. Schaphorst, A. Jean, O. Schwarzkopf, P. Lablanquie, L. Andric, A. Huetz, J. Mazeau, and V. Schmidt, J. Phys. B **29**, 1901 (1996).

[3] J. Viefhaus, G. Snell, R. Hentges, M. Wiedenhöft, F. Heiser, O. Gessner, and U. Becker, Phys. Rev. Lett. **80**, 1618 (1998).

[4] S. Rioual, B. Rouvellou, L. Avaldi, G. Battera, R. Camilloni, G. Stefani, and G. Turri, Phys. Rev. A **61**, 044702 (2000).

[5] F.H. Read, J. Phys. B: At. Mol. Phys. **10**, L207 (1977).

[6] J.P. van den Brink, J. van Eck, and H.G.M. Heideman, Phys. Rev. Lett. **61**, 2106 (1988).

[7] J.P. van den Brink, G. Niehnhuis, J. van Eck, and H.G.M. Heideman, J. Phys. B: At. Mol. Opt. Phys. **22**, 3501 (1989).

[8] J.A. de Gouw, J. van Eck, A.C. Peters, J. van der Weg, and H.G.M. Heideman, Phys. Rev. Lett. **71**, 2875 (1993).

[9] J.A. de Gouw, J. van Eck, A.Q. Wollrabe, J. van der Weg, and H.G.M. Heideman, Phys. Rev. A **50**, 4013 (1994).

[10] R.R. Blyth et al., J. Electron Spectrosc. Relat. Phenom. **101-103**, 959 (1999).

[11] G.C. King, M. Tronc, F.H. Read, and R.C. Bradford, J. Phys. B: At. Mol. Phys. **10**, 2479 (1977).

[12] N. Shanti, P.C. Deshmukh, and S.T. Manson, Phys. Rev. A **37**, 4720 (1988).

[13] D.W. Lindle et al., Phys. Rev. A **33**, 319 (1986).

[14] B. Kämmerling, B. Krassig, O. Schwarzkopf, J.P. Ribeiro and V. Schmidt, J. Phys. B: At. Mol. Opt. Phys. **25**, L5 (1992).

[15] L. Vegh, Phys. Rev. A **50**, 4036 (1994).

[16] P. Selles, J. Mazeau, P. Lablanquie, L. Malegat, and A. Huetz, J. Phys. B: At. Mol. Opt. Phys. **31**, L353 (1998).

[17] J. Tulkki, N. M. Kabachnik, and H. Aksela, Phys. Rev. A **48**, 1277 (1993).

[18] J Cooper and R.N. Zare, J. chem. Phys. **48**, 942 (1968); J. Cooper and R.N. Zare, J. Chem. Phys. **49**, 4252 (1968).

[19] E.G Berezhko, N.M. Kabachnik and V.S. Rostvosky, J. Phys. B: At. Mol. Phys. **11**, 1749 (1978); E.G Berezhko and N.M. Kabachnik, J. Phys. B: At. Mol. Phys. **12**, 2993 (1979)

[20] G. Snell, B. Langer, A.T. Young, and N. Berrah, Phys. Rev. A **66**, 022701 (2002).

Inst. Phys. Conf. Ser. No. 183
Paper presented at Int. Conf. Electron and Photon Impact Ionization and Related Topics, Louvain-la-Neuve, 2004

Correlations in the Auger process under electron impact ionisation of Argon

F. Catoire*, C. Dal Cappello+, A. Lahmam-Bennani* and A. Duguet*

* Laboratoire des Collisions Atomiques et Moléculaires, LCAM, (UMR 8625) and Fédération Lumière et Matière (FR 2764), Bât. 351, Université Paris XI, 91405 Orsay Cedex, France
+ Institut de physique, LPMC, 1 Bd Arago, 57078 Metz Cedex 3, France

Abstract. Unprecedented (e,3-1e) coincidence angular distributions for the double ionisation of argon where the detected pair of electrons has the same kinetic energy (205 eV) as the Auger electron issued from a 2p-ionisation and the subsequent electronic relaxation have been measured at three different impact energies. The experimental data are compared with preliminary theoretical results from a first Born - 3CW model, and possible dominant ionisation mechanisms are tentatively identified in each case.

1. Introduction

In this work we investigate the angular distribution of the Auger emission produced when an incident electron ionises an argon atom in an inner shell, leaving the resulting ion in an excited state which then decays *via* electronic relaxation. The reaction can be written as:

$$e_0 + Ar \rightarrow e_a + e_b + Ar^{+*}$$
$$Ar^{+*} \rightarrow Ar^{++} + e_c$$

where e_0, e_a, e_b, e_c are the incident and the three outgoing electrons, respectively.
The emission of the Auger electron is generally assumed to be isotropic, tough it has been shown in [1] that this is not necessarily the case. In order to investigate this process, we used our double toroidal spectrometer [2] to perform (e,(3-1)e) experiments where two out of the three outgoing electrons are detected in coincidence, but under the particular condition of having *the same kinetic energy (205 eV) as the Auger electron.* This is *a priori* expected to give rise to strong interference and/or exchange effects, which should considerably modify the angular distributions. Moreover, the coincidence angular distributions of the Auger electrons were measured for various incident energies to get an insight into the mechanisms responsible for the Auger process under the given kinematics.

The Auger process under investigation produces the $L_{23}M_{23}M_{23}$ manifold of Auger lines. However, due to the low energy resolution used in the experiments ($\Delta E \approx 3eV$), these transitions cannot be resolved.

In the following sections, we first describe the experimental results. Then preliminary results from a first order model that we are presently developing, where the Auger process is assumed to be a two step process, will be presented and qualitatively compared to the experiments.

2. Experimental procedure

A simple diagram of the experimental arrangement is sketched in Fig. 1. An incident electron (e_0) double ionizes an argon atom and is scattered (e_a) in the direction θ_a , while the two ejected electrons (labeled b and c) emerge in the directions θ_b and θ_c. All angles are counted positively counterclockwise, starting from the incident electron direction, with $0° \leq \theta_b \leq 180°$ and $180° \leq \theta_c \leq 360°$.

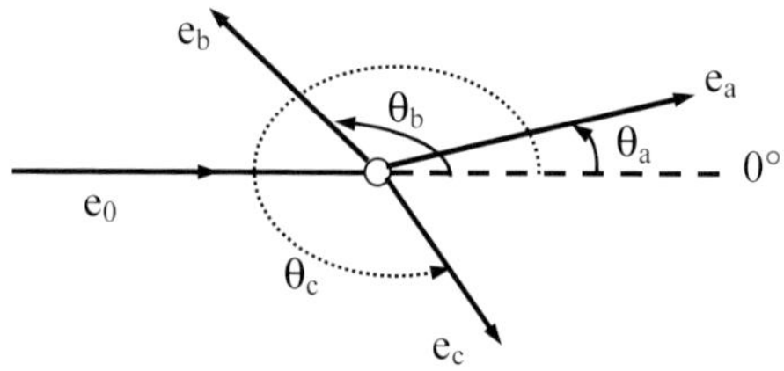

Figure 1 : Schematic diagram of the experimental arrangement

Two of the three final state electrons (labeled b and c for convenience) are energy and multi-angle analysed in the collision plane (**k**$_0$, **k**$_a$) using a double toroidal analyser and are detected in coincidence, see [2] for more details. After subtraction of the false coincidence signal the four-fold differential cross section distribution for the Auger process under investigation is reconstructed. The two detected electrons are chosen to be of the same energy as that of the Auger electron, i.e. 205 eV.

The incident energy is fixed at three different values: E_0 = 956 eV, 658 eV and 466 eV. This means from the energy balance that the third, unobserved electron has an energy of 503 eV, 205 eV and 13 eV, respectively. Finally, we note that, since the two observed electrons are completely indistinguishable, the cross section distribution must be symmetric with respect to the incident direction ($\theta_b = 360 - \theta_c$). This was indeed found to be the case in the experiments, with the deviation from symmetry never exceeding 15%. Hence, the data presented below are a symmetric average of the data collected in both half-planes with respect to the incident beam.

We now describe the different distributions obtained experimentally.

2.1. $E_0 = 956\ eV$

In this first case, the energy of the unobserved electron (labeled a) is the largest one, 503 eV. Hence, this electron might be considered as the scattered one, according to the commonly accepted nomenclature. In the intensity distribution displayed in Fig. 2, we observe mainly two peaks : a forward peak with respect to the incident direction at about the angles (60°; 300°), and a backward peak at about the angles (100°; 260°), both peaks lying on the symmetry line corresponding to the incident direction. Moreover, two secondary peaks are also present out of the symmetry line at coordinates (60°; 240°) and (120°; 300°).

2.2. $E_0 = 658\ eV$

At this incident energy, the three outgoing electrons are "completely indistinguishable", and one is dealing here with the problem of three identical particles in a coulombic potential. This means that one has to take into account the exchange of particles for this specific reaction as described in Section 3.

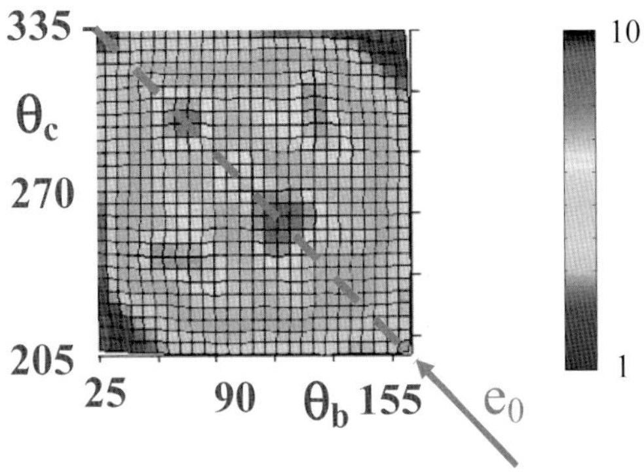

Figure 2 : Coincidence intensity distribution for the detection of two electrons of equal energy, 205 eV, plotted versus θ_b and θ_c ; $E_0 = 956$ eV and $E_a = 503$ eV. The intensity scale is linear and is given by the color scale to the right.

In figure 3, we can observe one main backward, broad peak at roughly (120°; 240°), and two secondary ones at (60°; 240°) and (120°; 300°). The secondary peaks are approximately at the same positions as for $E_0 = 956$ eV. In contrast, the main one corresponds to the remarkable situation where both ejected electrons are going out in the backward direction. From a symmetry argument based on the full identity of the three electrons, this probably means that the unobserved one is emitted in a forward direction, close to the incident one. Finally, we note the presence of a minimum of intensity (or 'a hole') at coordinates (70°; 290°). The presence of this hole will tentatively be explained in the following Section.

Another notable result is that the coincidence rate is about 2 to 3 times larger than in the cases where $E_0 = 466$ eV or $E_0 = 956$ eV. In a similar work performed over a much more limited angular range, Doering *et al* [3] reported a smaller or even a minimum cross section in the particular case where the three outgoing electrons have the same energy. The present results do not show such a behaviour, but rather an enhancement of the cross section for this equal energy sharing situation. However, one must note that the geometry of the Doering *et al* experiment (non coplanar) is very different from the geometry of the present one.

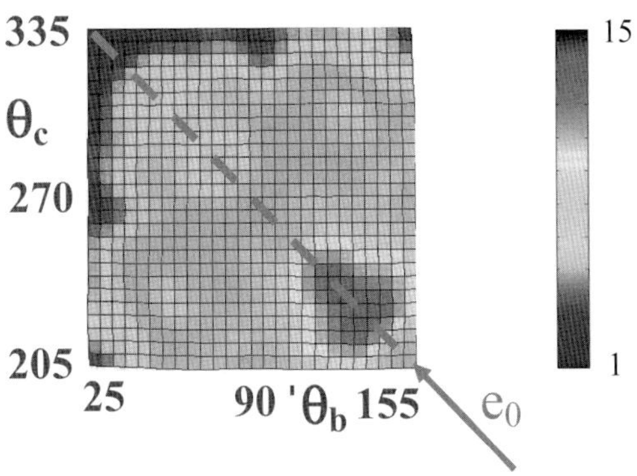

Figure 3 : Same as in Figure 1, except $E_0 = 658$ eV and $E_a = 205$ eV.

2.3. $E_0 = 466$ eV

At this incident energy, the data in Figure 4 are strongly dominated by the main forward peak at the position (50°; 310°). The two secondary peaks are still appearing at roughly the same position as in previous cases.

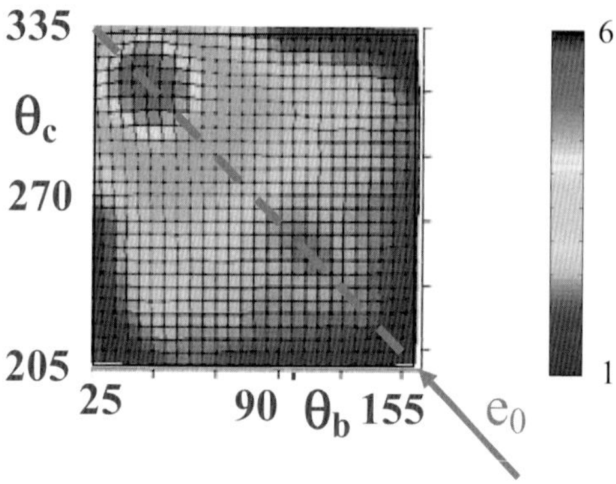

Figure 4 : Same as in Figure 1, except $E_0 = 466$ eV and $E_a = 13$ eV.

2.4. Discussion

We first make two remarks:

(i) Generally speaking, for a given final state, it is not possible to say whether the double ionisation is the result of direct double ionisation (DDI) of the 3p outer shell or double ionisation with an Auger process. However, if the DDI dominates over the Auger process (or inversely), then the structure of the experimental results can be attributed to the dominant process alone.

(ii) In the three cases the observed cross section distributions show secondary peaks at approximately the same position, hence quasi-independent of the incident energy. This can be brought together with the assumption that the Auger electron emission occurs in two *independent* steps, where the angular distribution related to the second step, i.e. to the target relaxation has no memory of the first inner shell hole creation and hence is independent of the incident energy. Therefore, under this hypothesis the secondary peaks could be attributed to Auger electrons.

It is possible to give a simple tentative explanation of the forward peak appearing at (50°; 310°) for $E_0 = 466$ eV and at (60°, 300°) for $E_0 = 956$ eV using classical mechanics as justified by the high electron energies involved. Let us assume a reaction where an incident electron (0) collides with a system made up of two stationary electrons (b and c). We can write the equations of energy and momentum conservation, but this system cannot be solved in the general case. However, this simplifies in our case because, first we have $k_b = k_c$, and second we may assume that the two ejected electrons emerge at the same angle with respect to the incident direction but in opposite half-planes, i.e. $\theta_b = 360 - \theta_c$ (this assumption is consistent with the experimental results). From momentum conservation it thus follows that the angle θ_a of the unobserved electron is 0 or π. We may then express θ_b as:

$$cos(\theta_b) = \frac{k_0 \pm k_a}{2k_b}$$

For $E_0 = 466$ eV and $\theta_a = 0°$, we find $\theta_b = 51°$. This value is close to the experimental one of ≈ 50°. Similarly, for $E_0 = 956$ eV, we find $\theta_b = 73°$, not far from the experimental value of ≈ 60°.

The situation is more complicated for $E_0 = 658$ eV. Indeed, interference between the two processes (DDI and Auger process) does occur as for the other energies, but in addition interference of exchange should be particularly strong in this specific case due to the full indistinguishability. Furthermore, from an experimental point of view, all possible pairs among the three outgoing electrons can be detected. These are the pairs Auger - ejected, Auger – scattered and ejected – scattered. The two first pairs necessarily arise from the Auger process. The pair ejected – scattered cannot be detected when $E_0 = 956$ eV and 466 eV, since one of these two electrons does not have the correct energy, 205 eV. In contrast, for the specific case $E_0 = 658$ eV, the detection of this pair is possible. But this is simply an (e,2e) process corresponding to the single ionisation of the 2p shell. The contribution of this pair adds incoherently to those from the other two pairs to build up the measured (e,3-1e) cross section in Fig. 3. Such an (e,2e) distribution was modelled using a DWBA approximation [4] and is presented in Fig. 5. Qualitatively, we can observe noticeable similarities between this distribution and the experimental one in Fig. 3 for the positions of the backwards peaks and the 'holes'. (Nonetheless, the presence of symmetric peaks at (40°; 220°) in the experiments is not reproduced by the theory). Such similarities suggest that the measured distribution is mostly determined by the (e,2e) contribution.

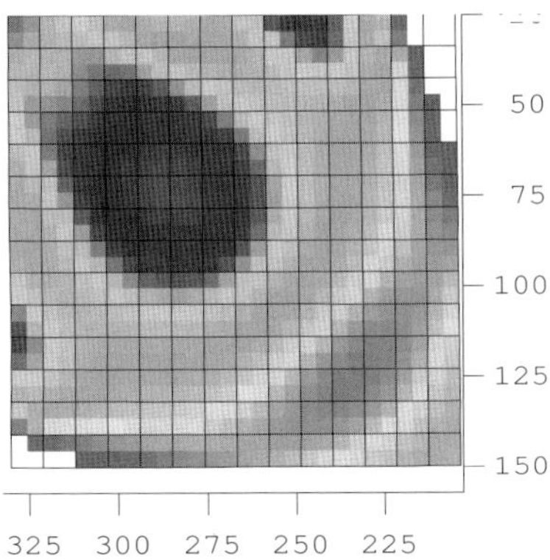

Figure 5 : (e,2e) DWBA intensity distribution versus θ_b and θ_c, for $E_0 = 658$ eV and $E_a = 205$ eV, as in figure 3.

3. Theoretical approach

In this section, we describe the model we are currently building to describe the Auger process and the DDI. It should be noted that the treatment of an atom with an inner shell hole that subsequently decays to yield an Auger electron has been addressed by Végh and Macek [5] in the context of photoionisation. Here, for simplicity, the DDI is assumed to be solely due to a shake off (SO) process. The SO process is described in the First Born Approximation (FBA). By using the well-known frozen-core approximation we reduce the 18-electron-target problem to a two-electron-target problem [6, 7], whereas the final state (the two ejected electrons and the scattered electron) is described by three coulomb waves (3CW), each coulomb wave representing the interaction between the electron and the ion. We also introduce three Gamow factors representing the repulsion between two electrons in the final state. In this model we also take into account the interference due to electron exchange.

The Auger process is assumed to be a two step model where the second step is described according to perturbation theory and using the Breit operator. The first inner-shell ionisation is described by the FBA-2CW model with the use of the frozen-core approximation which allows us to reduce this problem to a one-electron-target problem. Since the Auger electron feels the field of a doubly charged ion, it is also described by a coulomb wave with an effective charge Z = 2.

After the DDI, the Ar^{++} ion has fifteen states, and we cannot distinguish for a given final state wether this doubly ionised state is a result of the SO process or the Auger process. Therefore, the cross section is written as the coherent sum of both amplitudes. If we write t^{Au} for the probability amplitude for the Auger process and t^{SO} for the SO process, the differential cross section is given by: $\sigma \propto |t^{Au} + t^{SO}|^2$.

3.1. Description of the Auger process

The Auger process is described as a two step mechanism, where the first step is the single ionisation from the 2p shell, as illustrated in Figure 6. This first step is in fact an (e,2e) process which leaves the singly charged ion in an excited state. Using the FBA and 2CW model, this transition amplitude is written as follows:

$$t^{e2e} = \left\langle \Psi^{+} \left| \frac{-1}{r_a} + \frac{1}{r_{ab}} \right| \Psi_i \right\rangle$$

where a denotes the incident electron (and scattered) and b the ejected one.

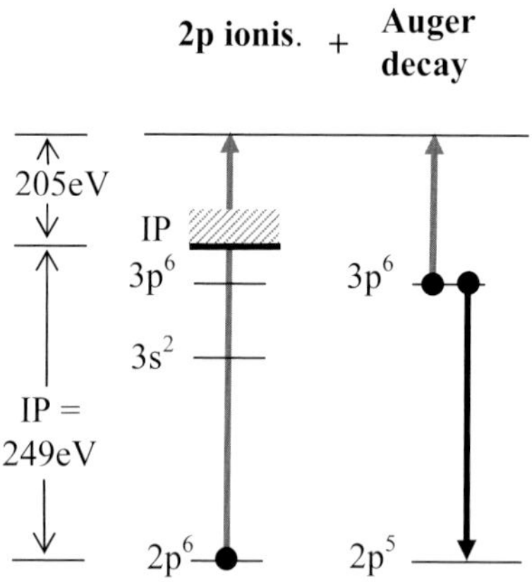

Figure 6 : Schematic diagram of the Auger transition described as a two step process.

The initial state is described by a one active electron wavefunction (Clementi) [8] and a plane wave for the incident electron, and the final state is a product of two coulomb waves given by :

$$\langle \vec{r}_b, \vec{r}_a | \Psi^{+} \rangle = g(\eta) \prod_{j=a}^{b} \frac{1}{(2\pi)^{3/2}} e^{\frac{\pi}{2}\alpha_j} \Gamma(1 + i\alpha_j) e^{i\vec{k}_j \cdot \vec{r}_j} {}_1F_1(-i\alpha_j, 1, -i(k_j r_j + \vec{k}_j \cdot \vec{r}_j))$$

where $\alpha_j = \dfrac{2}{k_j}$ $\eta = \dfrac{1}{k_{ab}}$ and $g(\eta) = e^{\frac{\pi}{2}\eta} \Gamma(1 - i\eta)$

Here, $|g(\eta)|^2$ is the Gamow factor which describes the coulomb interaction between electrons a and b. This transition amplitude has no analytical expression and we use the Roy, Roy and Sil [9] expression that we have modified for the Ar initial state wavefunction.

The subsequent step, i.e. the Auger transition is written as follows:

$$t^{A} = \left\langle \Psi_f \left| \frac{1}{r_{cd}} \right| \Psi^+ \right\rangle$$

where now $\left|\Psi^+\right\rangle$ describes the two electrons (c and d) of the Ar^+ ion involved in this reaction. Here, d is the electron that makes the transition from the 3p shell to the 2p shell. Finally, $\left|\Psi_f\right\rangle$ is the final state of the Auger process, so it is described as a product of an electron in a 2p shell and one in the continuum (the Auger one) modelled by a coulomb wave. We also introduce two other Gamow factors for the repulsion between a and c and b and c.

Consequently, the global Auger transition is:

$$t^{Au} = t^{A} t^{e2e} = \left\langle \Psi_f \left| \frac{1}{r_{cd}} \right| \Psi^+ \right\rangle \left\langle \Psi^+ \left| \frac{-1}{r_a} + \frac{1}{r_{ab}} \right| \Psi_i \right\rangle$$

Furthermore, for a given final state, several Auger transitions can occur. All these transitions are added in the calculation thanks to a coherent sum of these amplitudes in the cross section.

3.2. Description of the SO process

The SO process is described in the FBA theory, where the transition amplitude for this reaction is written as follows:

$$t^{SO} = \left\langle \Psi_f \left| \frac{-2}{r_a} + \frac{1}{r_{ab}} + \frac{1}{r_{ac}} \right| \Psi_i \right\rangle$$

The cross section for this process is given by: $\sigma^{SO} = (2\pi)^4 \frac{k_s k_b k_c}{k_a} \left| t^{SO} \right|^2$

In this model one has to use a two active electron wavefunction, where the incident electron is described by a product of a plane wave (incident electron) and a product of the wavefunction for each electron of the 3p shell of Argon (Clementi [8]).

The final state is described by a product of three coulomb waves:

$$\left\langle \vec{r}_c, \vec{r}_b, \vec{r}_a \middle| \Psi^+ \right\rangle = g(\eta) \prod_{j=a}^{c} \frac{1}{(2\pi)^{3/2}} e^{\frac{\pi}{2}\alpha_j} \Gamma(1 + i\alpha_j) e^{i\vec{k}_j . \vec{r}_j} {}_1F_1(-i\alpha_j, 1, -i(k_j r_j + \vec{k}_j . \vec{r}_j))$$

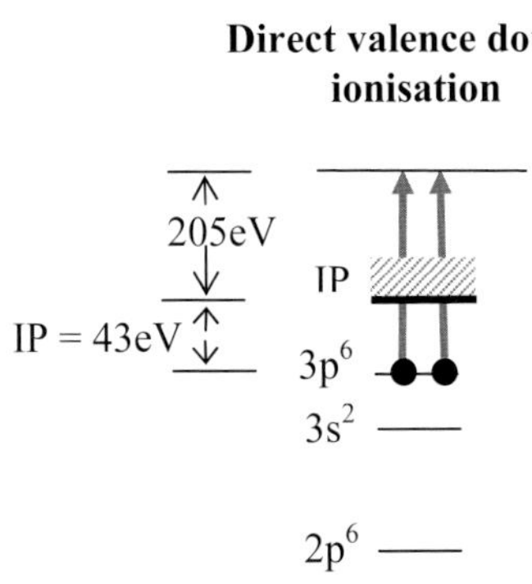

Figure 7 : Schematic diagram of the SO process.

Symbols used are the same as the ones used for (e,2e). Of course, the amplitude has no analytical expression, as was the case for (e,2e), so we have used the Roy, Roy and Sil [9] expression again.

The diagram of the SO process is given in Figure 7. This particular transition is a one step process.

3.3. *Interference due to exchange*

Due to the fact that the particles have identical energies, it is necessary to take into account their exchange. Consequently, the three outgoing electrons have to be described in the same way. That is why, in the final state, all electrons are coulombic waves. As in the paper of Schulz [10], it is possible that the incident electron after the ionisation does not become the scattered one, but what is usually called the ejected electron. Of course, this description has no real meaning because the electrons are indistinguishable. To sum up this result, the three electrons have to be eigenvectors of the spin matrix. The wavefunction of Argon has to be an eigenvector of the operator L^2 of this atom (described by two active electrons). Finally, the wavefunction of the system built up of three electrons has to be antisymmetric with respect to exchange of electrons. Note that the combination of two electrons in Ar gives fifteen different states, one for ${}^1S_0$, five for ${}^1D_0$ and nine for ${}^3P$. For each transition, we have to take the correct eigenvector built with the constraints described before, and all transition amplitudes will be expressed by the following terms:

$$f_{xy} = g_0^f(\vec{r}_a)g_1^f(\vec{r}_b)g_2^f(\vec{r}_c)\,(V)\,g_0^i(\vec{r}_a)g_x^i(\vec{r}_b)g_y^i(\vec{r}_c)$$
$$f_{xyi} = g_0^f(\vec{r}_a)g_1^f(\vec{r}_c)g_2^f(\vec{r}_b)\,(V)\,g_0^i(\vec{r}_a)g_x^i(\vec{r}_b)g_y^i(\vec{r}_c)$$
$$g_{xy} = g_0^f(\vec{r}_b)g_1^f(\vec{r}_a)g_2^f(\vec{r}_c)\,(V)\,g_0^i(\vec{r}_a)g_x^i(\vec{r}_b)g_y^i(\vec{r}_c)$$
$$g_{xyi} = g_0^f(\vec{r}_c)g_1^f(\vec{r}_a)g_2^f(\vec{r}_b)\,(V)\,g_0^i(\vec{r}_a)g_x^i(\vec{r}_b)g_y^i(\vec{r}_c)$$
$$h_{xy} = g_0^f(\vec{r}_b)g_1^f(\vec{r}_c)g_2^f(\vec{r}_a)\,(V)\,g_0^i(\vec{r}_a)g_x^i(\vec{r}_b)g_y^i(\vec{r}_c)$$
$$h_{xyi} = g_0^f(\vec{r}_c)g_1^f(\vec{r}_b)g_2^f(\vec{r}_a)\,(V)\,g_0^i(\vec{r}_a)g_x^i(\vec{r}_b)g_y^i(\vec{r}_c)$$

where the initial quantum numbers of the electrons of the initial state are x and y. Of course, if x = y the direct and indirect amplitudes are equal (as in the case of Helium). V is the interaction potential that we have seen before for the SO process.

For the Auger process, the inclusion of exchange is much more complicated than for the SO process. Indeed, the Auger process deals with the problem of four active electrons. Additionally, the global amplitude is the product of the amplitude of the first step with the potential V_1 and the amplitude of the second step with the potential V_2. Globally, in order to include this exchange for given initial and final states, these states are chosen among the eigenvectors found in the previous paragraph for SO. Thus, it will be possible to sum coherently the amplitude of the Auger process and the amplitude of the SO process including interference of exchange. At present, calculation of the amplitude for the SO process with exchange has been implemented. Work is proceeding on the inclusion of the Auger process in the calculations.

3.4. *Results*

In the absence of complete (e,3-1e) calculations for Ar which are still in progress, we present in figure 8 the (e,3e) theoretical results that we obtained with $E_0 = 466$ eV, $E_b = E_c = 205$ eV and a scattering angle of 20°. Results obtained at various other scattering angles show that the cross section distribution does not change significantly. In particular, the main peak at about (50°, 310°) is always present at roughly the same position for all scattering angles.

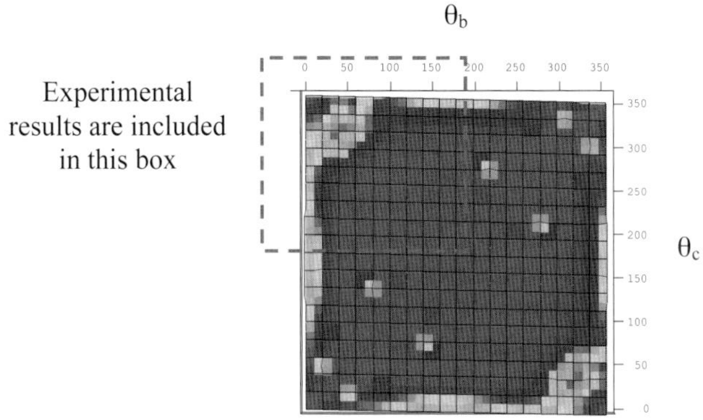

Figure 8 : (e,3e) intensity from the 3CW model with exchange, at E_0=466 eV, E_a=13 eV and θ_a=20°.

Furthermore, the wings around (180°,0°) and (0°,180°) have, in all cases, a smaller magnitude than the main peak. Hence, one can expect that the (e,3-1e) cross section which will be obtained by integrating the (e,3e) one should show the same behaviour : assuming that the main process is the SO, a dominant peak is expected to be present at the position (50°,310°), which is consistent with the experimental results in fig. 3. Of course, we do not expect the results in figure 8 to reproduce the secondary peaks present in the experiments (fig. 3) as they are *a priori* attributed to the Auger process.

For the other incident energies, it is not possible to draw a similar conclusion, since at higher energies the Auger process is probably dominant and hence the experimental results cannot be reproduced by the above described calculation, which only includes the SO process.

4. Conclusion and future work

This work gives a preliminary report of an ongoing, more complete study of the Auger process under electron impact of Argon for different incident energies. Experimentally, the reported (e,3-1e) results will be complemented by full (e,3e) experiments which will become feasible using a new toroidal analyser presently under test and which should soon be operational. On the theoretical side, we need to properly include the Auger process in the cross section calculation. Besides, integration over the solid angle of the unobserved electron is required. This time consuming calculation is in progress.

Nonetheless, some conclusions can already be drawn from the reported measurements and calculations:

(i) in the case of the lowest incident energy, the main process contributing to the observed intensity seems to be the DDI from the outer 3p shell, which is surprisingly well described by the SO process.

(ii) a simple classical picture involving billiard ball collisions seems to yield at least qualitatively the correct position of the forward peak observed at E_0 = 466 and 956 eV. This peak is not observed for the other incident energy of 658 eV.

(iii) in the intermediate case where $E_0 = 658$ eV, the main contribution to the angular distribution of the cross section seems to be the (e,2e) first ionisation step of the Auger process. This hypothesis is supported by the qualitatively reasonable agreement observed at this energy between the DWBA results and the experimental results.
(iv) the presence of wings in all the distributions of figures 2, 3 and 4 at the positions (60°; 240°) and (120°; 300°) could be attributed to the Auger process. Indeed, the global amplitude of the Auger process is the product of two transition amplitudes (for the two step model that we consider), one corresponding to the first inner-shell (e,2e) ionisation and the second one to the electronic recombination. The (e,3-1e) experiments involve an integration of this global cross section over the solid angles of the unobserved electron. If the resulting "ejected" electron and Auger electron angular distributions are relatively insensitive to the incident energy, this could explain the fact that these wings do not change their position as a function of incident energy.

References

[1] Berezhko E G, Kabachnik N M and Sizov V V 1978 J. Phys B. **11**, 1819
[2] Duguet A, Lahmam-Bennani A, Lecas M, and ElMarji B, 1998 Rev. Sci. Instrum. **69**, 3524
[3] Doering J P, Coplan M A, Cooper J W, and Moore J H, 1990 Phys. Rev. A **41**, 535
[4] Cavanagh S, and Lohmann B, 1997 J. Phys B. **30**, L231
[5] Végh L and Macek J H 1994 Phys. Rev. A **50** 4031
[6] Hda H, Dal Cappello C and Langlois J 1994 Z. Phys. D **29**, 25
[7] Dal Cappello C, Hda H and Roy A C 1995 Phys. Rev. A **51**, 5
[8] Clementi E and Roetti C, 1974 At. Data Nucl. Data Tables **14** 177
[9] Roy A, Roy K and Sil N C, 1980 J. Phys. B **13** 3443
[10] Schulz M, J. Phys. B 1973 6, 2580

Inst. Phys. Conf. Ser. No. 183
Paper presented at Int. Conf. Electron and Photon Impact Ionization and Related Topics, Louvain-la-Neuve, 2004

Doppler-free resonant Auger Raman spectroscopy on atoms and molecules at SPring-8

A. De Fanis†, Y. Tamenori†, M. Kitajima‡, H. Tanaka‡ and K. Ueda♯

† Japan Synchrotron Radiation Research Institute, Sayo-gun, Hyogo 679-5198, Japan

‡ Department of Physics, Sophia University,Tokyo 102-8544, Japan

♯ Institute of Multidisciplinary Research for Advanced Materials, Tohoku University, Sendai 980-8577, Japan

Abstract.
At the high resolution electron spectroscopy end station of BL27 at SPring-8 in Japan we replaced the conventional gas-cell with a Doppler free gas-jet source. Such a device was constructed in such a way that the gaseous molecules, or atoms, travel perpendicularly to the direction of observation of the electron energy analyzer, so that the Doppler broadening is removed. This apparatus has been tested and proved successfully for resonant Auger emission of Ne $1s^{-1}3p$ and H_2O O $1s^{-1}2b_2$ and O $1s^{-1}4a_1$.
In the experiment on Ne, electrons of about 811 eV of energy are observed with a width of 63 meV. Such narrow width allows to resolve terms of inner-valence excited states of the series Ne^+ $2s2p^5(^{1,3}P)np$ ^2S, ^2P, ^2D; and also to extract the natural lifetime widths of states that can undergo further Auger decay.
This Doppler-free technique was also applied in H_2O, to study the Auger decay from the O $1s^{-1}2b_2$ and O $1s^{-1}4a_1$ core-excited states. The Doppler-free system allows to resolve the vibrational structure in the final Auger state, thus to investigate details of nuclear dynamics in the core excited state.

1. Introduction

One of the ultimate factor limiting the amount of information that can be extracted from electron spectroscopy experiments is the overall resolution. Several components contribute to the resolution. According to the specific situation only one, or a few, of these contributions dominate. A particularly interesting situation is when one experimental technique is greatly improved to the level that a specific contribution to the experimental resolution is reduced from dominant to negligible. For example, in the case of normal Auger spectroscopy the resolution of the electron energy analyzer is the dominant contribution to the overall width if this is larger than the natural lifetime width of the core-hole ionized states, Γ, which is nowadays no longer the case. In resonant Auger spectroscopy, where the decaying state is a core-excited state of the neutral system rather than a core-ionized atom [1], two opposite limits exist. If the photon width is much greater than the natural width, a convolution of the natural lifetime width and the analyzer width determines the width of the observed line. If the photon width is much smaller than the natural width, a convolution of the photon width and the analyzer width determine the width of the observed line. In fact, this situation is now possible thanks to the great improvement of soft X-ray monochromators in modern synchrotron radiation facilities experienced during the last decade: it is now possible to promote a core electron to a valence unoccupied orbital with the photon width smaller than the lifetime width of the excited state. This particularly interesting situation has been exploited very extensively in the last decade to access new information on spectroscopy of singly charged ions [2, 3, 4, 5] or dynamics of core-excited molecules [6, 7, 8].

If one includes the photon width, the electron energy analyzer width, and the Doppler broadening, the total broadening caused by experimental conditions is written as :

$$\Delta_{total} = \Delta_{h\nu} \oplus \Delta_{KE} \oplus \Delta_{Doppler}, \tag{1}$$

with the Doppler broadening that can be approximated as

$$\Delta_{Doppler} = 0.722 \times \sqrt{KE \times T/M}, \tag{2}$$

with the kinetic energy of the electron observed in eV, and T and M the temperature and mass of the gaseous sample in Kelvin and atomic unit, and $\Delta_{h\nu}$ in eV. If each contribution add a Gaussian distribution, then:

$$\Delta_{total}^2 = \Delta_{h\nu}^2 + \Delta_{KE}^2 + \Delta_{Doppler}^2. \tag{3}$$

When investigating the 1s resonant Auger decay of a light molecule (or atom) with a first row element, like H_2O O $1s^{-1}2b_2$ or Ne $1s^{-1}3p$ with very narrow photon width and electron energy width, the Doppler broadening can be the dominant one [2, 3] . We present results on Ne $1s^{-1}3p$ and H_2O O $1s^{-1}2b_2$ and O $1s^{-1}4a_1$ resonant Auger spectroscopy recorded with a gas-jet Doppler free source installed at BL27SU at SPring-8 [9].

2. Experimental apparatus

The main parts of the Doppler-free electron spectroscopy apparatus are the beam device (MBJ-01 by MB Scientific) and a 200-mm radius hemispherical electron spectrometer (SES-2002 by Gammadata-Scienta). The beam device has an orifice of 1 mm x 8 mm, covered by a 1 mm thick micro channel plate. The channel diameter is 10 μm, and the open area ratio is approximately 70 %. The surface facing the excitation region is covered with a conductive layer of graphite with suitable work function. An electrode structure surrounds the orifice, which allows precision modelling of the electrostatic potential in the target volume. The electrode voltages, as well as the voltage applied to the device body, can be externally adjusted to optimize the energy resolution. A multi-channel array permits a much higher target pressure than a single tube device. The inlet pressure of an effusive beam source is limited roughly to 1 Torr, i.e., the level where the mean free path regarding intermolecular collisions equals the channel length. The achievable target pressure scales as the square root of the number of channels when all other parameters are kept constant. The channel length to diameter ratio (i.e., 100 in our device) determines the beam divergence angle. The Doppler width is proportional to the width of the velocity distribution of the atoms or molecules parallel to the electron emission direction. The reduction of the Doppler width scales linearly with the ratio between the effective exciting beam diameter (typically 1 mm), and the distance to the beam (typically 5 mm). Reduction of more than 10 times can be easily achieved with minimal intensity loss.

The molecular beam is directed downward and enters a 2000 l/sec turbo-molecular pump. The lens axis of the analyzer is in the horizontal direction so that the analyzer can see the molecular beam in the perpendicular direction; the lens axis is also perpendicular to the photon beam. The entrance slit of the spectrometer is parallel to the target volume. In this alignment, the Doppler effect can be suppressed, because the transverse velocity component of the molecular beam is negligible.

The Doppler free apparatus was installed on the c-branch of beamline 27SU [9, 10, 11] at the synchrotron source SPring-8 in Japan. The radiation source of this beamline is a figure-8 undulator [12]. The direction of the linear polarization vector for the first-order harmonic light generated by this undulator is horizontal whereas that of the 0.5th-order harmonic light is vertical. Thus, one can perform angle-resolved electron spectroscopy with an electron spectrometer fixed in the horizontal direction, switching the direction of the polarization vector by changing only the undulator gap.

3. Experimental procedure and data analysis

For the experiment on neon we set the distance between the molecular beam source (orifice) and the source volume where the photon beam and the molecular beam cross to about 3 mm. We recorded sequentially and repeatedly resonant Raman Auger spectra for decaying from the Ne $1s^{-1}3p$ excited states at 867.12 eV [13] to Ne^+ $2p^4(^1D)3p$, $2sp^5(^{1,3}P)np$ and $2s^02p^6(^1S)np$ ^2S states (n=3, 4). The kinetic energy scale was calibrated by comparing the energy of the Ne $1s^{-1}3p \rightarrow Ne^+$ $2p^4(^1D)3p$ ^2P, ^2F and ^2D with the energies reported in [2], the accuracy is better than 100 meV. The acquisition time for each energy region was set so that the total intensity for each region was nearly equal. The polarization was repeatedly changed from 0° to 90° and vice versa. Measurements at 0° to 90° were calibrated in relative intensity by assuming that the integrated $2p^4(^1D)3p$ band is isotropic. This assumption in fact agrees with earlier measurements and ab-initio calculations, as well as with the expectation from the spectator model for resonant Auger [2, 4]. The polarization was > 0.95 [14]. Small drifts of the photon energy during time, caused by change of heat load on the beamline optics, were compensated by aligning the spectra in kinetic energy using the $2p^4(^1D)3p$ ^2F band as a calibrant.

The Ne spectra are χ^2-fitted with Lorentzian profiles convoluted with a the instrumental broadening. The Lorentzian profiles are as follows :

$$L(E) = \frac{A}{2\pi} \frac{\Gamma}{(E-E_0)^2 + \Gamma^2/4}, \tag{4}$$

where A is proportional to the cross section, E_0 is the center of the distribution and Γ is the full-width at half maximum (FWHM) and corresponds to the natural lifetime width. The instrumental profile was chosen as the sum of two Gaussians to represent the Doppler-free

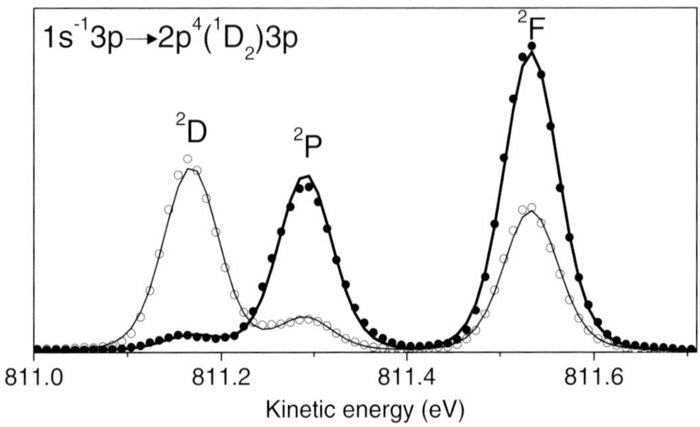

Figure 1. Resonant Auger spectra Ne $1s^{-1}3p \rightarrow Ne^{+}\ 2p^4(^1D_2)3p$. Dots (black and white) : experiments (0° and 90°); lines : fit

and the remaining Doppler-broadened components, respectively. Each of the Gaussians is as follows :

$$G_i(E) = \frac{2\sqrt{\log 2}}{\sigma_i\sqrt{\pi}} \exp\left[-\left(\frac{E-E_0}{\sigma_i}\right)^2 4\log 2\right], \tag{5}$$

where σ_i is the FWHM and E_0 the center of the distribution. Then the convolution is written as follows:

$$C(E) = \int_{-\infty}^{+\infty} L(x)\left[\frac{1}{1+c}G_1(E-x) + \frac{c}{1+c}G_2(E-x)\right]dx, \tag{6}$$

where c controls the relative intensities of the two profiles.

For the experiment on H_2O we set the distance between the molecular beam source and the source volume to about 8 mm, in order to suppress further the Doppler effects. The monochromator slits and the electron analyzer widths were set to a combined values expected to be 52 meV.

4. Results and discussion

4.1. Ne $1s^{-1}3p$

In figure 1 we present the Doppler-free resonant Auger spectra of Ne $1s^{-1}3p \rightarrow Ne^{+}\ 2p^4(^1D)3p$ ^2P, ^2D and ^2F. The final states $Ne^{+}\ 2p^4(^1D)3p$ are not energetically allowed to undergo further Auger decay. The energy separations between the doublet components are negligible [4, 5]. A separate measurement confirmed that a profile of each of these Auger lines coincides with that of the Ne 2s and 2p photolines within the experimental uncertainties. Thus, these spectra were used for extracting the experimental broadening in (6). The result is that the two Gaussians have FWHMs of 63 and 101 meV, and the ratio for their integrated areas is 1:0.45 : the peak-heights are different by nearly a factor of 4.

In figure 2 we present the spectra for the $2s^12p^5(^{1,3}P)np$ (n=3, 4) states. These inner-valence excited states can experience further Auger decay, which further broadens these transitions. The fit in the figure was carried out with the experimental profile in (6) extracted from

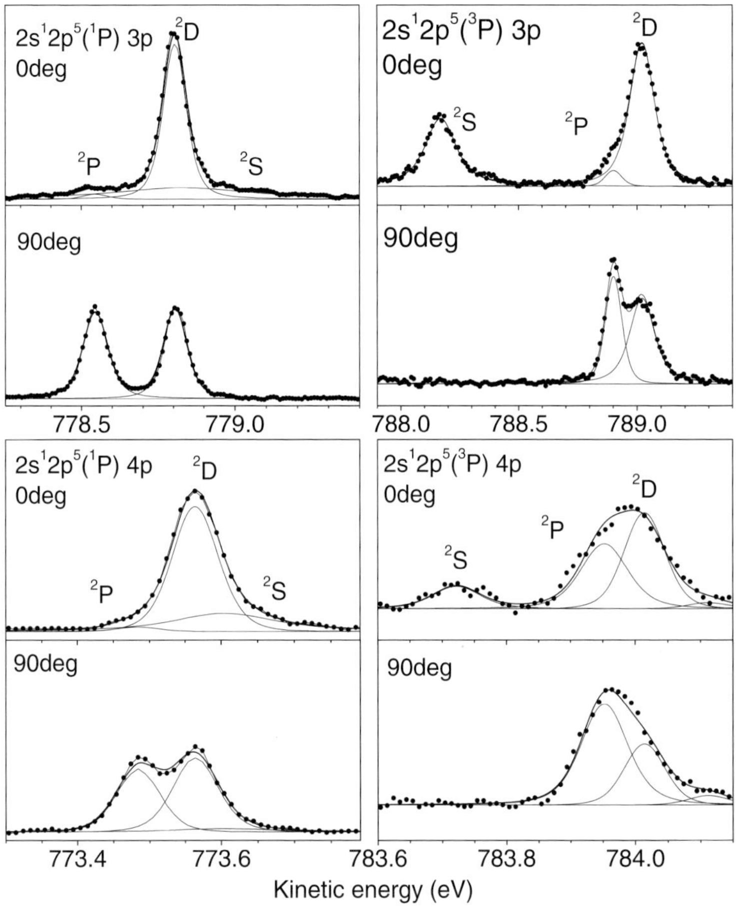

Figure 2. Resonant Auger spectra Ne $1s^{-1}3p \rightarrow Ne^{+}$ $2s^{1}2p^{5}(^{1,3}P_{2})np$ (n=3, 4). Dots : experiments (0° and 90°); thick and thin lines : fit and individual Lorentzian components.

the analysis of the spectra in figure 1. The assignment for the transitions to the $2s^{1}2p^{5}(^{1,3}P)3p$ states were made by comparing energy separations, relative intensities and angular distribution with those obtained by ab-initio calculation [4, 5], these measured and calculated values are listed in table 1. The assignments for the transitions to the $2s^{1}2p^{5}(^{1,3}P)4p$ states were made assuming that the branching ratios and the anisotropy parameters are similar to those for the transitions to $2s^{1}2p^{5}(^{1,3}P)3p$ states, as expected from the shake-up model.

The extracted widths for the $2s^{1}2p^{5}(^{1,3}P)$ np states are given in table 2. The widths extracted from this experiment are in agrement with those from satellites photoemission of [15]. The width of the $2s^{1}2p^{5}(^{1,3}P)np$ ^{2}S states shows better agreement with the elaborate calculations by Sinanis *et al.* [16] rather than with the predictions of Armen and Larkins [17] obtained in a single-configuration Hartree-Fock approximation: the low-energy intermultiplet transitions are very sensitive to electron correlations and can be only described if the

Table 1. Transition energies E_k, angular anisotropy parameters β and branching ratios (for each parent state) for the resonant Auger decay to the $2s^1 2p^5(^{1,3}P)np$ and $2s^0 2p^6\ ^2S$ np (n=3, 4) states of Ne^+.

final state	E_k (eV) expt.	E_k (eV) theor.	β expt.	β theor.	br. ratio expt.	br. ratio theor.
$2s^1 2p^5(^1P)3p\ ^2S$	778.79	776.43	$1.7^{+0.3}_{-0.4}$	2.000	0.13±0.03	0.099
$2s^1 2p^5(^1P)3p\ ^2P$	778.54	776.40	-0.94±0.06	-0.996	0.34±0.02	0.340
$2s^1 2p^5(^1P)3p\ ^2D$	778.81	776.66	0.20 ± 0.06	0.200	0.54±0.03	0.561
$2s^1 2p^5(^3P)3p\ ^2S$	788.16	786.51	1.5±0.1	1.998	0.18±0.02	0.153
$2s^1 2p^5(^3P)3p\ ^2P$	788.90	787.52	-0.91±0.06	-0.928	0.23± 0.03	0.324
$2s^1 2p^5(^3P)3p\ ^2D$	789.01	787.64	0.06±0.05	0.156	0.59± 0.03	0.523
$2s^1 2p^5(^1P)4p\ ^2S$	773.60	-	0.96±0.28	-	0.14±0.03	-
$2s^1 2p^5(^1P)4p\ ^2P$	773.48	-	-0.92±0.06	-	0.29±0.02	-
$2s^1 2p^5(^1P)4p\ ^2D$	773.56	-	0.15±0.06	-	0.57±0.04	-
$2s^1 2p^5(^3P)4p\ ^2S$	783.72	-	1.97±0.40	-	0.04±0.01	-
$2s^1 2p^5(^3P)4p\ ^2P$	783.95	-	-0.43±0.09	-	0.57±0.08	-
$2s^1 2p^5(^3P)4p\ ^2D$	784.01	-	0.09±0.20	-	0.35±0.20	-
$2s^0 2p^6(^1S)3p\ ^2P$	755.17	-	0.05 ± 0.14	-	-	-
$2s^0 2p^6(^1S)4p\ ^2P$	750.04	-	0.07 ± 0.18	-	-	-

Table 2. Widths, in meV, for the $2s^1 2p^5(^{1,3}P)np$ and $2s^0 2p^6(^1S)np$ (n=3, 4) states of Ne^+.

final state	Experiment This work	Experiment ref. [15]	Theory ref. [17]	Theory ref. [16]
$2s^1 2p^5(^1P)3p\ ^2S$	530 ± 50	410± 50	687	510
$2s^1 2p^5(^1P)3p\ ^2P$	42 ± 3	-	20.7	-
$2s^1 2p^5(^1P)3p\ ^2D$	34 ± 4	-	40.2	-
$2s^1 2p^5(^3P)3p\ ^2S$	120 ± 10	110 ± 40	18.8	122
$2s^1 2p^5(^3P)3p\ ^2P$	19 ± 5	-	10.3	-
$2s^1 2p^5(^3P)3p\ ^2D$	80 ± 10	-	62.3	-
$2s^1 2p^5(^1P)4p\ ^2S$	135± 20	-	-	-
$2s^1 2p^5(^1P)4p\ ^2P$	0±5	-	-	-
$2s^1 2p^5(^1P)4p\ ^2D$	9±4	-	-	-
$2s^1 2p^5(^3P)4p\ ^2S$	30± 15	-	-	-
$2s^0 2p^6(^1S)3p\ ^2P$	80± 5	-	-	-
$2s^0 2p^6(^1S)4p\ ^2P$	20± 5	-	-	-

latter are fully taken into account [16]. The widths of other multiplets ^2P and ^2D are much smaller. The measured values are rather close to the predictions of Armen and Larkins [17] as

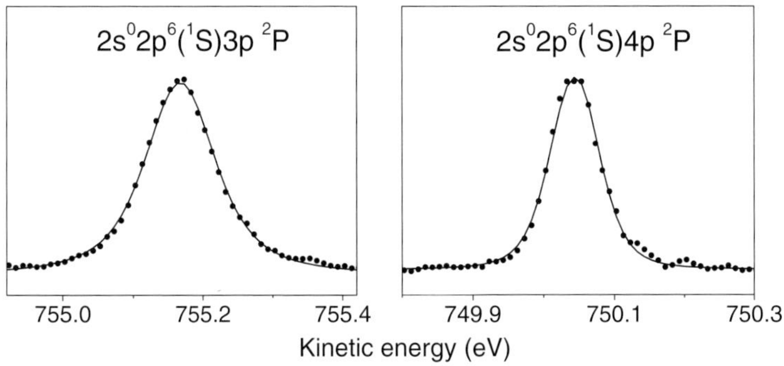

Figure 3. Resonant Auger spectra Ne $1s^{-1}3p \rightarrow Ne^+\ 2s^02p^6(^1S)np$. Dots and lines : experiments and fits.

obtained in a single-configuration Hartree-Fock approximation. Note, however, that the single-configuration approach fails to reproduce the experimental width of the ^2S states. An analysis of the contributions to the width of core $2s2p^5\ ^1P$ states from the inner-valence participator Auger transitions and the intermultiplet Auger transitions [17] shows that the former is dominant for the ^2P and ^2D lines while the latter determines the width of the ^2S line. Obviously, the low-energy intermultiplet transitions are much more sensitive to electron correlations and can be only described if the latter are fully taken into account [16].

Figure 3 presents portions of the electron spectra corresponding to the transitions to $Ne^+\ 2s^02p^6(^1S)3p\ ^2P$ and $2s^02p^6(^1S)4p\ ^2P$. Since it turned out that the values of β for these transitions are zero within the experimental uncertainties, we display the spectra in the angle-integrated form: I(0°)+2I(90°) where I(0°) and I(90°) are the spectra recorded for parallel and perpendicular polarizations. Now we focus on the n-dependence of all the widths extracted from the present measurements. The widths of $2s^12p^5(^{1,3}P)4p$ are smaller than those of $2s^12p^5(^{1,3}P)3p$ and those of $2s^02p^6(^1S)4p$ smaller than those for $2s^02p^6(^1S)3p$. This n dependence of the widths is a natural consequence that the np Rydberg electron participates in the electronic decay. In the electronic decay in which the Rydberg electron participates, the autoionization rate is roughly proportional to $(n^*)^{-3}$, where n^* is the effective principal quantum number of the Rydberg level [18].

4.2. H_2O O $1s^{-1}2b_2$ and O $1s^{-1}4a_1$

Figure 4 presents the electron spectra of H_2O in the binding energy interval of the valence-ionized states $X(1b_1^{-1})$, $A(3a_1^{-1})$ and $B(1b_2^{-1})$. The vibrational components are labelled (v_1,v_2) with v_1 and v_2 the quantum numbers for symmetric stretching and bending modes. The uppermost spectrum is recorded using the gas-cell whereas all others are recorded under Doppler-free conditions. The spectrum at the bottom was recorded with a He I lamp, with photon energy of 21.218 eV, and is presented here only for comparison, the others were recorded with monochromatic synchrotron radiation at BL27. For the spectra recorded with synchrotron radiation the photon width and the width of the electron analyzer were set to about 40 and 33 meV respectively. Unresolved rotational structure may provide additional contribution to the line broadening; the drift of the photon energy during the measurements may also add a small contribution to the line broadening.

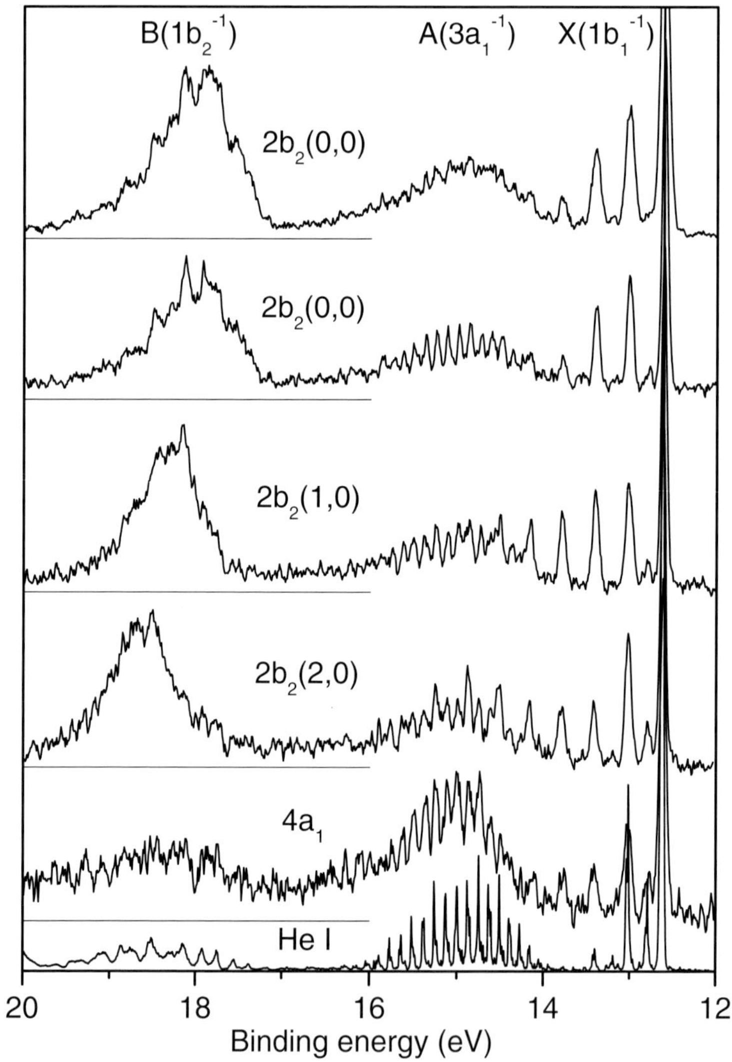

Figure 4. Comparison between resonant Auger electron spectra of H_2O with the direct photoemission (at the bottom). The top spectra was recorded with the gas-cell, all others under Doppler-free condition.

First, let us compare the spectra recorded at the O $1s^{-1}2b_2$ (0,0) excitation with the gas-cell and under Doppler-free condition. The X band is dominated by the (v,0) progression corresponding to the symmetric stretching with a spacing of 398 meV [21] and starting at 12.62 eV. In the Doppler-free spectrum is clear that the vibrational progression (v,1), starting at the energy of 12.78 eV is also excited. In the gas-cell spectrum the resolution is not sufficiently high to say whether the A band contains more than one vibrational progression, whereas in the

Doppler-free spectrum is clear that only the bending progression (0,v) with frequency of 108 meV [21] is present, the stretching mode has a frequency of 439 meV and is not present. The structure underneath the A band is probably the extended vibrational progressions that belong to the X band.

Let us now focus on the B band and compare the resonant spectra with the direct photoemission one. In both the gas-cell spectrum and the Doppler-free one, we cannot resolve the vibrations and thus cannot say to what extent the stretching and bending progressions are excited. The spacings of the stretching and bending modes in this band are 368 and 198 meV respectively [21]. Thus, if only one of the two modes was excited, we could see the progression. By comparing the spectra with the one recorded with the He I lamp it seems the reason the vibrational structure is not clearly resolved is because of multimode vibrational excitations in the final states. Note also that this B state may be predissociative for the vibrational components above 18 eV [20].

Let us now discuss the photon energy dependence of the Doppler-free resonant spectra. When the photon energy is increased in the O $1s^{-1}2b_2$ excitation the $(v_1, 0)$ progression in the X bands extends to higher vibrational components, and also the $(v_1, 1)$ progression becomes stronger. The centroid of the B band shifts towards higher binding energy, indicating that higher vibrational components are excited in the final Auger states. With the help of ab-initio calculation De Fanis *et al* [3] concluded that the reason why the vibrational structure in the B band is less resolved at higher energy is because the multimode vibrational excitation becomes increasingly more complex in the final Auger state with the increase in the vibrational excitation in the core-excited state. Even with the improved resolution here is still not possible to extract the intensity of each vibrational component.

At the O $1s^{-1}4a_1$ excitation only the A band is enhanced. The vibrational progression remains similar to the one in the direct photoemission, dominated by the bending mode, without presence of the stretching mode. This result is in contrast with Hjelte *et al* [19], where the limited resolution made the authors conclude more than one vibrational progressions were present.

A more detailed analysis of the vibrational structure in the spectra presented would require calculation of the potential energy surface of the excited state, and its decay. Though this is beyond the present work, the authors wish the present result will encourage theoreticians to perform such calculations.

References

[1] Armen G B, Aksela H, Åberg T and Aksela S 2000 *J. Phys. B: At. Mol. Phys.* **33** R49

[2] Shimizu Y, Yoshida H, Okada K, Muramatsu Y, Saito N, Ohashi H, Tamenori Y, Fritzsche S, Kabachnik N M, Tanaka H and Ueda K 2000 *J. Phys. B: At. Mol. Phys.* **33** L685

[3] De Fanis A, Saito N, Kitajima M, Shimizu Y, Okada K, Tanaka H, Koyano I and Ueda K 2001 *J. Phys. B: At. Mol. Phys.* **34** L377

[4] Ueda K, Kitajima M, De Fanis A, Tamenori Y, Yamaoka H, Shindo H, Furuta T, Tanaka T, Tanaka H, Yoshida H, Sankari R, Aksela S, Fritzsche S and Kabachnik NM 2003 *Phys. Rev. Lett.* **90** 153005

[5] De Fanis A, Tamenori Y, Kitajima M, Tanaka H and Ueda K 2004 *J. Electron. Spectrosc.* **136-140** 271

[6] De Fanis A, Nobusada K, Hjelte I, Saito N, Kitajima M, Okamoto M, Tanaka K, Yoshida H, Hiraya A, Koyano Y, Piancastelli M N and Ueda K 2002 *J. Phys. B: At. Mol. Phys.* **35** L23

[7] Ueda K 2003 *J. Phys. B: At. Mol. Phys.* **36** R1

[8] Ueda K, De Fanis A, Saito N, Machida M, Ku bozuka K, Chiba H, Muramatsu Y, Sato Y, Czasch A, Jagutzki O, Dörner R, Cassimi A, Kitajima M, Furuta T, Tanaka H, Sorensen S L, Okada K, Tanimoto S, Ikejiri K, Tamenori I, Ohashi H and Koyano I 2003 *Chem. Phys.* **289** 135

[9] Tamenori Y, Kitajima M, De Fanis A, Shindo H, Furuta T, Machida M, Nagoshi M, Ikejiri I, Yoshida H, Ohashi H, Koyano I, Tanaka H, Baltzer P and Ueda K 2002 *Proc. 19th Int. Conf. on X-ray and Inner-Shell Processes, Rome* (AIP) p 71

[10] Ohashi H, Ishiguro E, Tamenori Y, Okumura H, Hiraya A, Yoshida H, Senba Y, Okada K, Saito N, Suzuki I H, Ueda K, Ibuki T, Nagaoka S, Koyano I and Ishikawa T 2001 *Nucl. Instrum. Methods Phys. Res. A* **467-468** 529

[11] Ohashi H, Ishiguro E, Tamenori Y, Okumura H, Hiraya A, Yoshida H, Senba Y, Okada K, Saito N, Suzuki I H, Ueda K, Ibuki T, Nagaoka S, Koyano I and Ishikawa T 2001 *Nucl. Instrum. Methods Phys. Res. A* **467-468** 533

[12] Tanaka T and Kitamura H 1996 *J. Synchrotron Radiat.* **3** 47

[13] Coreno M, Avaldi M, Camilloni R, Prince K C, de Simone M, Karvonen J, Colle R and Simonucci S 1999 *Phys. Rev.* A **59** 2494

[14] Yoshida H, Senba Y, Morita M, Goya T, De Fanis A, Saito N, Ueda K, Tamenori Y and Ohashi 2004 *Proc. Synchrotron Radiation Instrumentation Conference* (AIP) p 267

[15] Pahler M, Caldwell C D, Schaphorst S J and Krause M O 1993 *J. Phys. B: At. Mol. Phys.* **26** 1617

[16] Sinanis C, Aspromallis G and Nicolaides C 1995 *J. Phys. B: At. Mol. Phys.* **28** L423

[17] Armen G B and Larkins F P 1991 *J. Phys. B: At. Mol. Phys.* **24** 741

[18] Fano U and Cooper J W 1968 *Rev. Mod. Phys* **40** 441

[19] Hjelte I, Piancastelli MN, Fink RF, Björneholm, Bässler M, Feifel R, Giertz A, Wang H, Wiesner K, Ausmess A, Miron C, Sorensen SL and Svensson S 2001 *Chem. Phys. Lett.* **334** 151

[20] Hjelte I, De Fanis A, Carravetta V, Saito N, Kitajima M, Tanaka H, Yoshida H, Hiraya A, Koyano I, Karlsson L, Svensson S, Ueda K and Piancastelli MN to be submitted.

[21] Reutt JE, Wang LS, Lee YT and Shirley DA 1986 *J. CHem. Phys.* **85** 6928

Inst. Phys. Conf. Ser. No. 183
Paper presented at Int. Conf. Electron and Photon Impact Ionization and Related Topics, Louvain-la-Neuve, 2004

Interference effects in ionization of H_2 targets by fast electron impact

C R Stia†, O A Fojón†‡, R D Rivarola†‡, J Hanssen||, O Kamalou§, D Martina§, J -Y Chesnel§ and F Frémont§

† Instituto de Física Rosario, CONICET-UNR, Av. Pellegrini 250, 2000 Rosario, Argentina

‡ Escuela de Ciencias Exactas y Naturales, Facultad de Ciencias Exactas, Ingeniería y Agrimensura, Universidad Nacional de Rosario, 2000 Rosario, Argentina

|| Institut de Physique, Laboratoire de Physique Moléculaire et des Collisions, Université de Metz, Technopôle 2000, 1 Bv. Arago, 57078 Metz Cedex 3, France

§ Centre Interdisciplinaire de Recherche Ions Lasers, Unité Mixte CEA-CNRS-EnsiCaen-Université de Caen Basse Normandie, 6 Bd du Mal Juin, F-14050, Caen Cedex France

Abstract. In the present work, we investigate the ionization of hydrogen molecules by fast electron impact. In particular, we study the interference phenomenon which may appear as consequence of the coherent emission from the two indistinguishable centers of the molecule. Interference patterns are predicted in the ejected electron spectra corresponding to multiple differential cross sections. Moreover, recent experimental interference evidence is presented for the case of single ionization of the D_2 targets by electron impact.

1. Introduction

During the last decades, a large number of theoretical and experimental studies have treated the electron-impact ionization of atomic targets providing a general understanding of these processes (see, for example, reference [1]). On the contrary, the research on ionization of molecules has advanced in a slower way due to some extra complications appearing in this case. For instance, the rotational and vibrational states of the molecular targets cannot be determined during a general collision experiment. On the theoretical side, both the extra degrees of freedom and the multicentric nature of the molecular targets must be taken into account in any realistic description of the process.

In the present work, we investigate the single ionization of diatomic molecules at high enough incident energies for different kinematical arrangements. Particular attention is paid to the interference effects which may appear in the spectra of the ionized electrons. This effect is due to the two-center character of the initial bound state in which the target electrons are shared by both nuclei of the molecule [2].

The interference patterns were predicted in the case of photoionization of H_2 and H_2^+ molecules [2, 3, 4] and recently measured for electron emission by the impact of fast krypton ions [5, 6] on H_2 targets. These works have promoted an active research on this topic [7, 8, 9]. But it is only recently that these effects have been also studied both theoretically [10] and experimentally [11] in (e,2e) reactions involving H_2 and D_2 molecules as targets, respectively. The results of these works will be summarized in the present report.

This paper is organized as follows. After a brief outline of the theoretical model used in the calculations, we present cross sections results corresponding to the ionization of H_2 molecules by fast electron impact. In particular, we study the structures appearing in the multiple differential cross sections as a function of the ejected electron angle. It is found that the oscillatory patterns depend strongly on the molecular orientation. In addition, double differential cross sections for (e,2e) reactions are also computed and compared with the recently measured data corresponding to D_2 targets [11] allowing a more detailed discussion about the interference phenomenon. For instance, constructive interferences are identified in the binary encounter region as well as in the low ejected electron velocities domain. Atomic units will be used except where otherwise stated.

2. Theory

We consider the single ionization process of hydrogen molecules following the impact of fast electrons, i.e.,

$$e^- + H_2(^1\Sigma_g^+) \rightarrow 2e^- + H_2^+(^2\Sigma_g^+). \tag{1}$$

At the high incident energies considered in this work, some approximations can be made in order to simplify the theoretical description [12, 13]. Briefly, the ionization process may be considered as a pure electronic transition by applying closure relations over all possible final rotational and vibrational states. Also, the molecule can be considered as frozen during all the reaction. In particular, we assume that the internuclear distance remains unchanged during the collision. Thus, only vertical electronic transitions at fixed equilibrium distance ρ_0 from the electronic ground state of the H_2 target to the electronic ground state of the residual H_2^+ molecular ion are considered in the calculations. Moreover, as only highly asymmetric arrangements are studied exchange effects may be neglected. Accordingly, the five-fold differential cross section (5DCS) can be written as

$$\sigma^{(5)} = \frac{d^5\sigma}{d\Omega_p d\Omega_e d\Omega_s d(k_e^2/2)} \cong 2(2\pi)^4 \frac{k_e k_s}{k_i} \left| t_{fi}^e(\boldsymbol{\rho}_0) \right|^2 \tag{2}$$

where the emitted electron is ejected with momentum $\mathbf{k}_e$ into the solid angle Ω_e with respect to the incidence direction defined by the momentum $\mathbf{k}_i$ of the incident electron, and the projectile is scattered with momentum $\mathbf{k}_s$ into the solid angle Ω_s. $\boldsymbol{\rho}_0$ denotes the equilibrium internuclear vector of the molecular target and $d\Omega_p$ its corresponding differential solid angle. In order to take into account the two electrons of the H_2 molecule, an extra factor 2 is included in the expression (2).

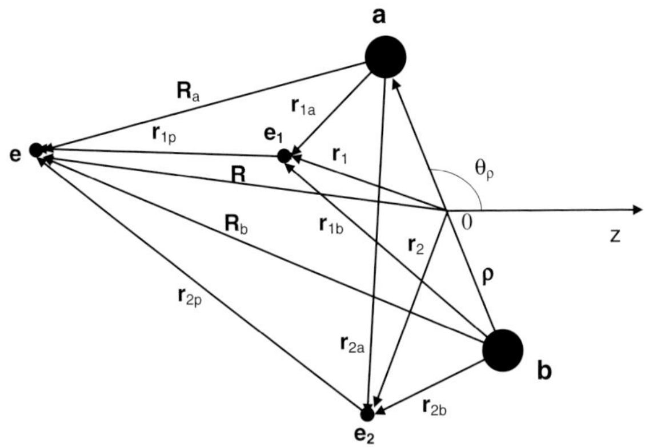

Figure 1. Coordinates used in the text.

The *prior* version of the electronic transition matrix element reads

$$t^e_{fi}(\rho_0) = \left\langle \Psi_f^-(\rho_0, \mathbf{R}, \mathbf{r}_1, \mathbf{r}_2)|V_i|\Psi_i(\rho_0, \mathbf{R}, \mathbf{r}_1, \mathbf{r}_2)\right\rangle \tag{3}$$

where Ψ_i is the non-perturbed electronic wave function in the initial channel and Ψ_f^- is the final exact electronic wave function with correct asymptotic conditions. In equation (3), V_i represents the perturbation in the initial channel given by

$$V_i = \frac{1}{r_{1p}} + \frac{1}{r_{2p}} - \frac{Z_T}{R_a} - \frac{Z_T}{R_b} \tag{4}$$

with $Z_T = 1$, the charge of the molecular nuclei. Coordinates are sketched in figure 1.

The initial wave function Ψ_i is represented as

$$\Psi_i = \frac{e^{i\mathbf{k}_i \cdot \mathbf{R}}}{(2\pi)^{3/2}} \Phi_i(\rho, \mathbf{r}_1, \mathbf{r}_2) \tag{5}$$

where the initial molecular bound state Φ_i is described here by a Heitler-London type wave function [14],

$$\Phi_i(\rho_0, \mathbf{r}_1, \mathbf{r}_2) = N_{HL}(\rho_0)\,\{e^{-\alpha^* r_{1a}} e^{-\alpha^* r_{2b}} + e^{-\alpha^* r_{1b}} e^{-\alpha^* r_{2a}}\} \tag{6}$$

with $\alpha^* = 1.166$, $\rho_0 = 1.406$ and N_{HL} the normalization function.

The choice of the approximate final electronic wave function is made within the framework of the molecular three-continuum approximation [15]. In this model, both the two-effective center description (TEC) [16] and the correlated three-body continuum wave function previously used in the atomic case [17] are employed in order to describe the final channel (see reference [15] for details). In this way, it may be shown that molecular amplitudes reduce to a coherent sum of two three-bodies amplitudes [10], where the three bodies are the incident electron, the scattered electron and one or the other molecular nucleus. The molecular three-continuum model [15] was used with success to describe absolute triple differential cross sections (TDCS) for single ionization of H_2 under the same energetic regime of the present interest [18]. Furthermore, this model can also be used to describe measured TDCS at intermediate incident energies [19].

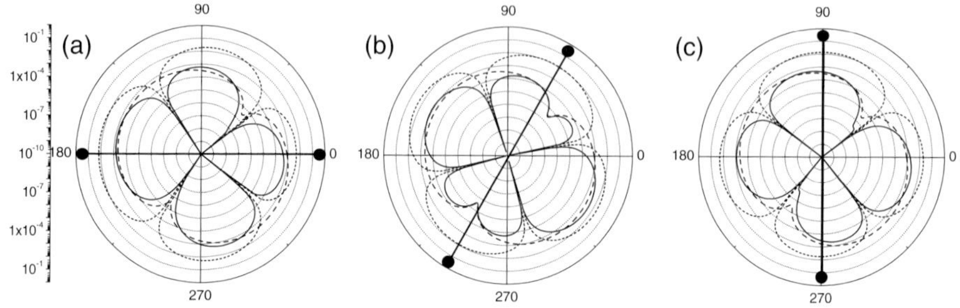

Figure 2. Five-fold differential cross sections for the electron impact ionization of hydrogen molecules as a function of the ejection angle. The impact and emission energies are $E_i = 4087$ eV and $E_e = 150$ eV, respectively. (a) $\theta_\rho = 0°$. (b) $\theta_\rho = 45°$. (c) $\theta_\rho = 90°$.

The exact final wave function may be then approximated as

$$\Psi_f^- \cong \frac{e^{i\mathbf{k}_s.\mathbf{R}}}{(2\pi)^{3/2}} \Phi_f(\rho, \mathbf{r}_2)\, \xi_c \tag{7}$$

where Φ_f is the wave function corresponding to the bound state of the residual H_2^+ molecular ion. It is represented in the present work by a simple linear combination of atomic orbitals,

$$\Phi_f(\rho, \mathbf{r}_2) = N_f(\rho)\, \{e^{-\alpha r_{2a}} + e^{-\alpha r_{2b}}\} \tag{8}$$

with $\alpha = 1.3918$ and $N_f(\rho)$ the normalization factor evaluated at ρ_0=1.406. The function ξ_c in equation (7) is chosen as

$$\xi_c(\mathbf{k}_e, \mathbf{k}_s, \mathbf{R}, \mathbf{r}_1, \mathbf{r}_2) = \frac{e^{i\mathbf{k}_e.\mathbf{r}_1}}{(2\pi)^{3/2}} C(\mathbf{k}_e, \mathbf{r}_{1j}, \gamma_e)\, C(\mathbf{k}_s, \mathbf{R}_j, \gamma_s)\, C(\mathbf{k}_{1p}, \mathbf{r}_{1p}, \gamma_{ep}) \tag{9}$$

where $j = a$ or b and the Sommerfeld parameters γ are given by

$$\gamma_s = -\frac{Z_T}{k_s} \qquad \gamma_e = -\frac{Z_T}{k_e} \qquad \gamma_{ep} = \frac{1}{2k_{1p}}. \tag{10}$$

Here, $\mathbf{k}_{1p} = \frac{1}{2}(\mathbf{k}_s - \mathbf{k}_e)$ is the momentum conjugate to $\mathbf{r}_{1p}$. So, the function ξ_c describes the mutual interactions between the three particles in the final channel of the reaction through the use of the Coulomb wave functions $C(\mathbf{k}, \mathbf{r}, \gamma)$ given by

$$C(\mathbf{k}, \mathbf{r}, \gamma) \;=\; \Gamma(1 - i\gamma)\, e^{-\pi\gamma/2}\, {}_1F_1[i\gamma; 1; -i(kr + \mathbf{k}.\mathbf{r})]. \tag{11}$$

Following Ref. [10], molecular 5DCS become

$$\sigma^{(5)} = \frac{d^5\sigma}{d\Omega_\rho d\Omega_e d\Omega_s d(k_e^2/2)} \cong 2\, \left[1 + \cos(\boldsymbol{\chi}.\boldsymbol{\rho}_0)\right]\, \sigma_A^{(3)} \tag{12}$$

where $\boldsymbol{\chi} = \mathbf{k}_e - \mathbf{K}$, and $\mathbf{K} = \mathbf{k}_i - \mathbf{k}_s$ is the momentum transferred to the ionized electron. The function $\sigma_A^{(3)}$ appearing in equation (12) represents a one-center TDCS corresponding to effective H atoms placed at the position of each one of the molecular centers. These effective H

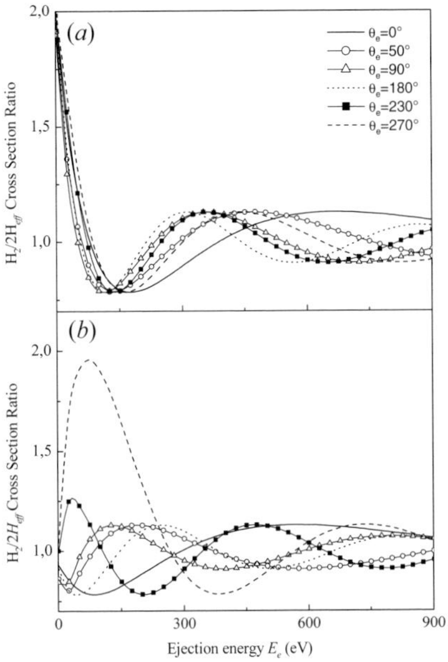

Figure 3. Normalized MBBK TDCS for the ionization of H_2 molecules as a function of the ejection energy for several fixed ejection angles. The incidence energy is $E_i = 4087$ eV while the scattering angle is (a) $\theta_s = 1°$ and (b) $\theta_s = 8°$.

atoms are computed by using an effective charge and the molecular bound energy corresponding to the Heitler-London initial wave function. The interference pattern due to the coherent emission from both molecular centers appears explicitly in equation (12).

Molecular TDCS corresponding to a coplanar geometry are obtained averaging equation (12) over all possible molecular orientations, i.e.,

$$\sigma^{(3)} = \frac{\mathrm{d}^3\sigma}{\mathrm{d}\Omega_e \mathrm{d}\Omega_s \mathrm{d}(k_e^2/2)} \cong 2\left[1+\frac{\sin(\chi\rho_0)}{\chi\rho_0}\right]\sigma_A^{(3)}. \tag{13}$$

Finally, double differential cross sections (DDCS) may be obtained from equation (13) after performing an additional integration over the solid angle corresponding to the scattered electron,

$$\sigma^{(2)} = \frac{\mathrm{d}^2\sigma}{\mathrm{d}\Omega_e \mathrm{d}E_e} \cong 2\int \mathrm{d}\Omega_s \left[1+\frac{\sin(\chi\rho_0)}{\chi\rho_0}\right]\sigma_A^{(3)} \tag{14}$$

Therefore, oscillations are predicted even in the case of the integrated cross sections.

3. Results and conclusions

In this section, multiple differential cross sections results are computed for electron-impact ionization of H_2 targets at high enough impact energies.

In figures 2(a)-(c), molecular 5DCS results as a function of the ejection angle θ_e are presented for some fixed orientations of the molecule. The impact and emission energies are

$E_i = 4087$ eV and $E_e = 150$ eV, respectively whereas the scattering angle is $\theta_s = 1^\circ$. Under the particular kinematical conditions considered in the figures, clear interference patterns are observed in the molecular spectra. Moreover, as the molecular orientation changes oscillations seem to rotate in a fixed-body frame, in the same way as it was observed for heavy ions as projectiles [20]. The present behavior in the 5DCS may be explained by considering the interference factor containing the cosine function (see equation (12)) which is also included in figures 2 after a proper scaling for a better comparison. It follows that the nodes in the molecular cross sections correspond to the nodes in the interference factor. Results corresponding to twice TDCS for effective H targets are also plotted in figures 2. At the asymmetric kinematical conditions considered here, atomic results show the expected double structure corresponding to the binary and recoil peaks [17].

Oscillations are also expected even after averaging over all possible molecular orientations as it could be inferred from equation (14). However, small interference effects may not be easily evidenced from the absolute cross sections [10]. So, instead of absolute molecular results, theoretical ratios between molecular TDCS and twice the cross sections corresponding to effective H atoms are plotted in figures 3 as a function of the ejection energy for several fixed ejection angles. The incident energy is $E_i = 4087$ eV and the scattering angles considered are $\theta_s = 1^\circ$ and $\theta_s = 8^\circ$. Normalized TDCS show marked oscillations depending on the electron ejection angle. In particular, at the smallest scattering angle considered in the figures, an increase in the frequency of the oscillations is found as the backward emission region is reached. The same behavior was observed in the case of the DDCS for heavy ions as projectiles [7]. In addition, maximum interference values occur at particular emission angles about 270° for which the Bethe condition is satisfied.

Finally, we present experimental evidence of the interferences in the case of the DDCS corresponding to the ionization of D_2 molecules by fast electron impact. Again, in order to make noticeable the interference effects, measured DDCS are divided by twice the DDCS corresponding to effective H atoms calculations. The resulting ratios (open circles) are represented in the figures 5 as a function of the ejected electron velocity for different ejection angle values. If interference effects were not present, the cross sections ratios at high incident energies would be approximately equal to unity. However, clear oscillations around unity are found except for the case of 70° in which experiments are distributed around the value 1.7. First-order TEC calculations (full lines) are also included in the figures showing a general good agreement with the experimental data.

In figures 5, we consider theoretical calculations in an extended domain of the emitted electron velocities. Maximum values are found for forward emission angles. These maxima come from the oscillatory term in the integrand of equation (14) which gives a maximum contribution for $\mathbf{k_e} = \mathbf{K}$, i.e., under binary encounter conditions. In particular for 70°, the theoretical ratios oscillate around a value of 1.7, in clear agreement with the measured values. For larger electron velocities, the theoretical ratios present damping oscillations around the unity as it occurs for the other ejection angles considered.

In summary, the presence of interference patterns due to the coherent emission from H_2 (D_2) targets was predicted theoretically and recently confirmed experimentally for the case of fast electrons as projectiles. In addition, constructive interferences were found in the binary encounter region supporting the existence of the phenomenon.

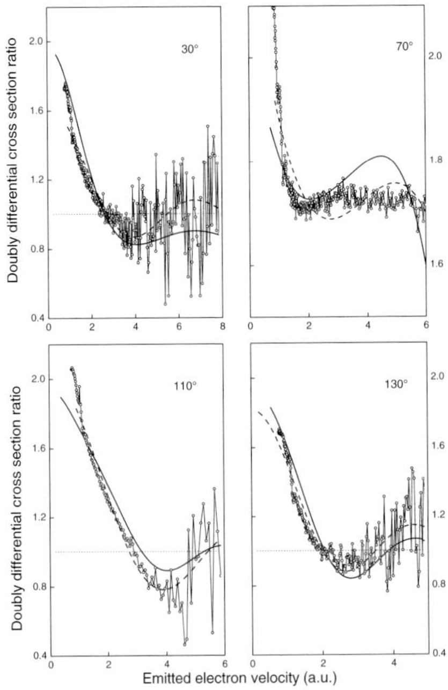

Figure 4. Normalized DDCS ratios corresponding to the single ioni zation of D_2 molecules as a function of the emitted electron velocity, for ejection angles of 30°, 70°, 110°, and 130°. The impact energy is $E_i = 2.4$ keV. Solid lines, theoretical first-order TEC calculations. Open circles, experimental data from Ref. [11].

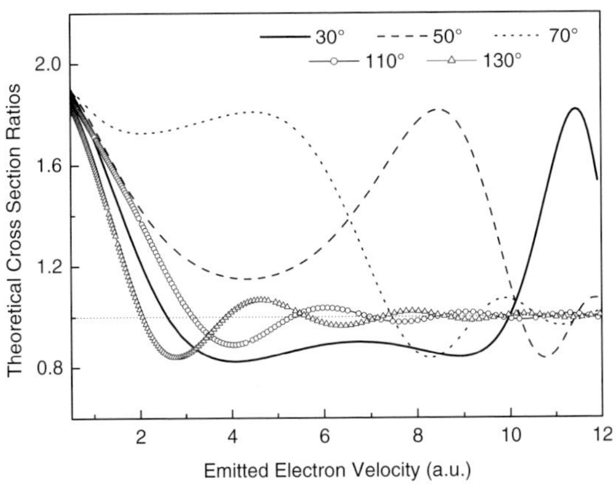

Figure 5. Normalized theoretical DDCS ratios for the ionization of H_2 molecules as a function of the ejection energy for several fixed ejection angles. Same kinematical conditions as in Fig. 4.

4. Acknowledgments

This work was partially supported by the French-Argentinean ECOS-Sud programme (N° A02E04). C.R.S., O.A.F. and R.D.R. acknowledge also support from the Consejo Nacional de Investigaciones Científicas y Técnicas de la República Argentina.

References

[1] Whelan C T, Walters H R J, Allan R J and Zhang X 1993 *(e,2e) and Related Processes* (Dordrecht: Kluwer)

[2] Cohen H D and Fano U 1966 *Phys. Rev.* **50** 30

[3] Walter M and Briggs J 1999 *J. Phys. B: At. Mol. Opt. Phys.* **32** 2487

[4] Fojón O A, Fernández J, Palacios A, Rivarola R D and Martín F 2004 *J. Phys. B: At. Mol. Opt. Phys.* **37** 3035

[5] Stolterfoht N, Sulik B, Hoffmann V, Skogvall B, Chesnel J Y, Rangama J, Frémont F, Hennecart D, Cassimi A, Husson X, Landers A L, Tanis J A, Galassi M E and Rivarola R D 2001 *Phys. Rev. Lett.* **87** 023201.

[6] Stolterfoht N, Sulik B, Gulyás L, Skogvall B, Chesnel J Y, Frémont F, Hennecart D, Cassimi A, Adoui L, Hossain S and Tanis J A, 2003 *Phys. Rev.* A. **67** 030702

[7] Galassi M E, Rivarola R D, Fainstein P D and Stolterfoht N 2002 *Phys. Rev.* A **66** 052705

[8] Hossain S, Alnaser A S, Landers A L, Pole D J, Knutson H, Robison A, Stamper B, Stolterfoht N and Tanis J A, 2003 in *Proceedings 11th Int. Conf. on the Physics of Highly Charged Ions, Nucl. Instrum. Methods* B **205** 484.

[9] Misra D, Kadhane U, Singh Y P, Tribedi L C, Fainstein P D and Richard P 2004 *Phys. Rev. Lett.* **92** 153201

[10] Stia C R, Fojón O A, Weck P, Hanssen J and Rivarola R D 2003 *J. Phys. B: At. Mol. Opt. Phys.* **36** L257

[11] Kamalou O, Martina D, Chesnel J -Y, Frémont F, Hanssen J, Stia C R, Fojón O A and Rivarola R D 2004 *Phys. Rev.* A *submitted*

[12] McCarthy I E 1973 *J. Phys. B: At. Mol. Phys.* **6** 2358

[13] Dey S, McCarthy I E, Teubner P J O and Weigold E 1975 *Phys. Rev. Lett.* **34** 782

[14] Wang S C 1928 *Phys. Rev.* **39** 579

[15] Stia C R, Fojón O A, Weck P F, Hanssen J, Joulakian B and Rivarola R D 2002 *Phys. Rev.* A **66** 052709

[16] Weck P, Fojón O A, Hanssen J, Joulakian B and Rivarola R D 2001 *Phys. Rev.* A **63** 042709

[17] Brauner M, Briggs J S and Klar H 1989 *J. Phys. B: At. Mol. Opt. Phys.* **22** 2265

[18] Chérid M, Lahmam-Bennani A, Zurales R W, Lucchese R R, Duguet A, Dal Cappello M C and Dal Cappello C 1989 *J. Phys. B: At. Mol. Opt. Phys.* **22** 3483

[19] Jung K, Schubert E, Paul D A L and Ehrhardt H 1975 *J. Phys. B: At. Mol. Phys.* **8** 1330

[20] Laurent G, Fainstein P D, Galassi M E, Rivarola R D Adoui L and Cassimi A 2002 *J. Phys. B: At. Mol. Opt. Phys.* **35** L495

Inst. Phys. Conf. Ser. No. 183
Paper presented at Int. Conf. Electron and Photon Impact Ionization and Related Topics, Louvain-la-Neuve, 2004

Quasi-elastic scattering of electrons at large momentum transfer

M. Vos†, G. Cooper‡ and C.A. Chatzidimitriou-Dreismann§

† Atomic and Molecular Physics Laboratories, Research School of Physical Sciences and Engineering, Australian National University, Canberra, ACT, 0200, Australia

‡ Department of Chemistry, McMaster University, Hamilton, Ontario, L8S 4M1, Canada

§ Institute of Chemistry, Stranski Laboratory, Technical University Berlin, Strasse des 17. Juni 112, D-10623 Berlin, Germany

Abstract. Precision measurements of the energy of keV electrons scattered 'elastically' over large angles from solid films or gas-phase molecules show that these electrons have lost some energy. In particular scattering from a system that contains hydrogen and carbon reveals two well-separated elastic peaks. The amount of energy lost in the deflection of the electron depends not only on the mass of the nucleus the electron scattered from, but also on the momentum of this nucleus. We make an estimate of the cross section for these elastic scattering processes. These cross sections do not reproduce the observed intensities very well, in particular the observed hydrogen intensity is for the solid film experiment smaller than expected. With increasing energy loss the time-scale of the collision process decreases. The decrease of the hydrogen cross section is possibly due to quantum correlation effects of the proton at the short timescale of the collision.

1. Introduction

It is well known that the region probed in a scattering experiment depends on the momentum transfer $\boldsymbol{q}$. The intensities originating from sources separated by $r >> 1/q$ add incoherently. If $r \leq 1/q$ one has to add the intensity coherently, leading to phenomena such as diffraction. At very large momentum transfer there is no structure left in the target, at a scale of $1/q$ and we have binary collisions from either a nucleus or an electron in the system. Thus in an electron scattering experiment at very large momentum transfer, one scatters from either a single electron or a single nucleus (i.e. a binary collision). If one scatters from an electron a significant fraction of the kinetic energy is transferred and these scattering events are studied in, e.g., $(e,2e)$ experiments. Here we restrict ourselves to electrons scattered from nuclei.

If one assumes that the target (mass m_t) is initially at rest (momentum $p_t = 0$), then a momentum transfer of an amount q implies that after the collision $p_t = q$ and hence it has an energy $q^2/2m_t$. This energy transfer should be reflected in a decrease in energy of the elastically scattered electron. The question is, which mass m_t should we consider: the mass of the whole solid or molecule or that of an individual nucleus. Experimentally it was shown by Boersch *et al* [1] as early as 1967 that at large momentum transfer the energy loss of the scattered electron is consistent with scattering from an individual nucleus.

If the target nucleus (mass m) has initially a momentum $\boldsymbol{p}_0$, then the energy transfer ω to the target is

$$\omega = \frac{(\boldsymbol{p}_o + \boldsymbol{q})^2}{2m} - \frac{p_o^2}{2m} = \frac{q^2}{2m} + \frac{\boldsymbol{p}_0 \cdot \boldsymbol{q}}{m} \tag{1}$$

Thus if the mass of the scatterer is known the measurement resolves a component of the momentum of the nucleus in a similar way as Compton scattering of an X-ray from an electron resolves a component of the electron momentum.

In spite of the work of Boersch *et al* it was generally believed that the energy transfer of an electron scattered from a nucleus was too small to be measured. It was realised that the momentum distribution of nuclei in materials (in particular the zero-point motion of the nucleus in its ground state) was determined by the strength of the bond, and hence was of fundamental interest. A technique was developed using neutrons to measure these momentum distributions, usually referred to as Neutron Compton Scattering (NCS) [2]. The theory behind these neutron scattering experiments is well established [2, 3, 4].

In the next section we present electron scattering data for both thin films and gas phase targets. The signal of electrons scattered from carbon and hydrogen atoms is well separated, and the effect of target motion is resolved. There appear to be problems reconciling the measured hydrogen intensity with the theoretically expected one for the thin film targets. We will explain that in the high-energy limit the electron scattering experiments are described by the same theory as the NCS experiments and compare our data with NCS experiments. Target properties obtained by neutron are in good agreement with those obtained from electron scattering experiments. For both probes a smaller hydrogen signal than expected was found. We will stress that these experiments probe the target on a very short time scale. It is argued that the shortfall in proton-derived intensity could be due to quantum correlations in the proton wave function at these short time scales.

2. Experimental Results

The electron scattering results for the polyethylene (PE) film were obtained using the ANU electron momentum spectrometer [5]. Thin, free-standing, samples ($\simeq$ 10 nm thick) were obtained using a method derived from the procedure described by Cranfill [6] and the one described by Godovsky and Magonov [7]. The polyethylene was dissolved in xylene at 100 °C (0.1% by weight). A microscope slide was partly submerged in this solution, and after it was thermally equilibrated, it was slowly (0.5 - 1 cm per minute) pulled out of the hot solution, leaving the slide coated with a thin PE film. After cooling down and drying the microscope slide was slowly submerged in distilled water with a 10-20° angle between the microscope slide and water surface. The PE film separates from the microscope slide and floats on the water surface. This film was then transferred to the sample holder, covering many holes.

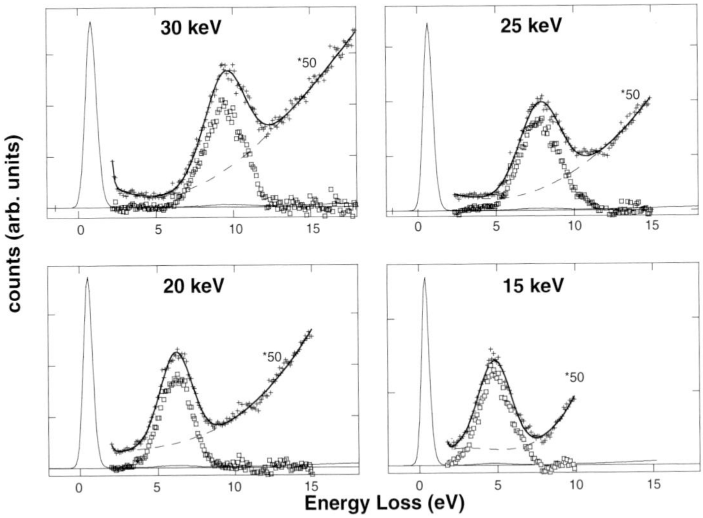

Figure 1. Energy loss spectra of polyethylene. Spectra are obtained for incoming electron energies as indicated. Besides the major elastic peak (attributed to electrons scattered from carbon) there is a second, small peak (shown as a + on the 50 times expanded scale) at larger energy loss values(attributed to electrons scattered from hydrogen). The separation of this peak relative to the main peak is proportional to the incoming energy. The background is shown as a dashed line, the fit as a full line and the background subtracted hydrogen peak as open squares.

The energy loss spectra, obtained for a scattering angle of 45° and electron energies between 15 and 30 keV, are shown in Fig. 1. As the precise zero energy-loss position of the spectrometer is difficult to determine we aligned the energy scale in such a way that the main peak position corresponds to the energy loss expected for scattering from a single carbon atom. The spectra all show a smaller peak at larger loss values superimposed on a background. The background is due to electrons that have scattered elastically from a carbon atom *and* have created an electronic excitation in the film. The shape of the background should resemble the energy loss spectra as measured in the forward direction. In particular this shape should not change with the energy of the probing electron. As the smaller peak is at an energy loss value proportional to the incoming energy, it can not be due to an elastic deflection plus electronic excitation. Indeed its position is what one would expect for quasi-elastic scattering from a

target	Θ deg.	E_0 (keV)	$\|q\|$ (a.u.)	ΔE_{obs} (eV)	ΔE_{calc} (eV)	Γ_{obs} (eV)	Γ_{calc} (eV)	$I_H{:}I_C$ obs.	$I_H{:}I_C$ calc.
PE	45	30	35.9	8.6	8.75	3.1	3.2	1:24.9	1:18
PE	45	25	32.9	7.1	7.35	3.1	2.9	1:24.4	1:18
PE	45	20	29.2	5.6	5.8	2.5	2.6	1:29	1:18
PE	45	15	25.6	4.4	4.45	2.4	2.3	1:25.1	1:18
CH_4	100	2.0	18.6	2.15	2.4	1.7	1.7	1:10.1	1:9

Table 1. A summary of the measured and calculated separation of hydrogen from the main peak, full width half maximum Γ of the hydrogen peak and the intensity ratio of the hydrogen peak I_H and main peak I_C

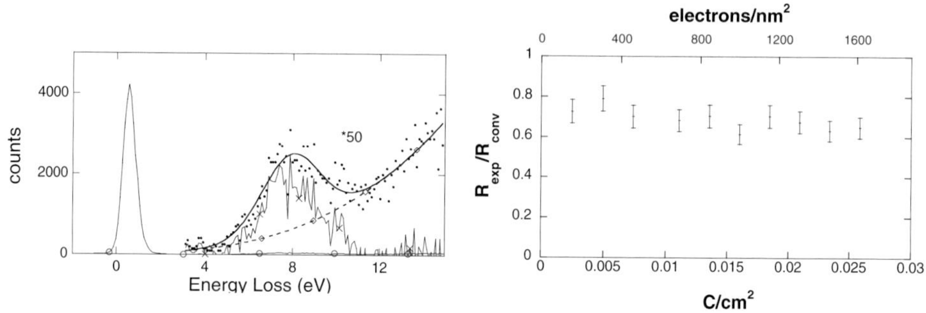

Figure 2. The left panel shows the quality of a polyethylene spectrum obtained in two minutes and the decomposition of the signal (dots) in background (dotted line) and proton signal (full line, magnified part. A series of two minute measurements was performed. The dependence of the R_{exp}/R_{conv} (measured hydrogen peak area relative to the expected peak area) as a function of accumulated beam charge is displayed in the right panel. The R_{exp}/R_{conv} ratio decreases slowly with dose. However this decrease is much too slow to explain the observed anomaly.

single proton. The spectra were fitted by two Gaussians and a polynomial background. The results of this fitting procedure is summarised in table 1.

The width of the proton-derived elastic peak is clearly much larger than the width of the carbon peak (and even the carbon peak is wider than the experimental resolution [8]). This extra width is due to Doppler broadening induced by the motion of the nucleus (see Eq. 1). Thus we have here a clear signature of the proton momentum distribution. For a proton in an isotropic, harmonic potential the momentum distribution is Gaussian, and the observed width is conveniently expressed in terms of the recoil energy and the mean kinetic energy of the atoms $\overline{E_k}$. The standard deviation σ of the observed Gaussian is given by [9]:

$$\sigma = \sqrt{\frac{4}{3}\overline{E_k}E_r}, \tag{2}$$

with E_r the mean recoil energy. As a first approximation all C-H bonds will be rather similar. If we take ethane (C_2H_6) as an example it has a calculated zero-point energy of 2 eV per molecule [10]. Half of this will be kinetic energy. Most of the kinetic energy will be in the lighter atoms, so we expect around 0.15 eV kinetic energy per proton. For hydrogen in hydrogenated amorphous carbon Mayers et al measure 146 meV using NCS [11], and they estimate on theoretical grounds energies between 156 and 183 meV for hydrogen in C-H bonds. Using Eq. 2 and a mean proton kinetic energy of 0.15 eV one obtains the estimate of the full-width half maximum given in table 1. The agreement between the observed and calculated values of the carbon-proton peak separation and the proton peak width is quite satisfactory.

For now we used a crude model for the cross section, assuming simply that at these momentum transfers the cross section is equal to the Rutherford cross section. From the known composition of polyethylene $(-CH_2-)_n$ and these cross sections we can make an estimate of the ratio of the carbon-hydrogen peak areas. We use a polynomial fit of the background as indicated with a dashed line in Fig. 1. Using the Rutherford cross section for electrons and the known composition of the target we can calculate the expected ratio of the hydrogen to carbon peak. We call this ratio R_{conv}. The observed ratio R_{obs} of the hydrogen peak relative to the

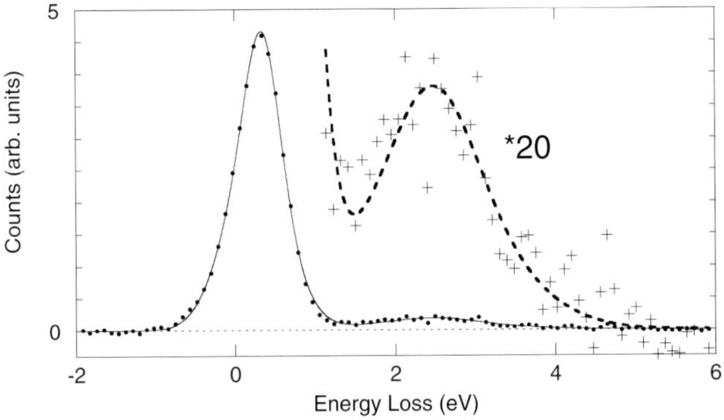

Figure 3. The spectrum of 2 keV electrons scattered over 100° from methane. Again two peaks are visible, consistent with electrons scattering from either C or H. The measurement is fitted with two Gaussians and a constant background (subtracted in the figure), as indicated by the full line (dashed line in the 20* magnification).

carbon peak is smaller than the calculated ratio R_{conv}. It appears as if some (about 25 %) of the protons are invisible.

In electron scattering experiments one has to be careful as the electron beam may cause a change in composition of the film. The impinging electrons can break a chemical bond and a hydrogen atom may subsequently desorb, leaving a carbon-rich film behind. In order to test if radiation damage is the cause of the smaller than expected hydrogen peak area, we studied the signal as a function of electron dose. By keeping the carbon-hydrogen peak separation fixed at the established value in the fitting routine, as well as the hydrogen and carbon peak width, we could obtain reasonably accurate peak areas for spectra taken with a low dose of electrons. We found a small dependency of R_{obs}/R_{conv} on electron exposure (see Fig.2), however this dependency was too weak to explain the observed discrepancy.

These results are intriguing. Are similar measurements possible using gas-phase targets? This would truly eliminate the problem of radiation damage, as in these effusion experiments the target is continuously replenished. Also the target density will generally be low enough that multiple scattering effects are absent, i.e. we do not have to subtract a poorly known background.

For this purpose we made preliminary measurements using the McVAHRES spectrometer at McMaster University[12]. The maximum incoming energy (2 keV) and scattering angle (100°) attainable in this spectrometer were chosen. The signal was weak, and a constant background had to be subtracted. The results are shown in Fig. 3. Again the main peak was aligned with the expected energy loss for electrons scattered from carbon under these conditions (0.2 eV). There is indeed again a broader satellite now at the energy loss of just over 2 eV, but now there is n o background increasing with energy loss. The spectra could be fitted using two Gaussians and a constant background term. The fitting results are summarised in table 1 as well. The peak width and separation is in good agreement with the calculated one. Preliminary indications are that the carbon to hydrogen peak area ratio is closer to the theoretical expected one than in the polyethylene case.

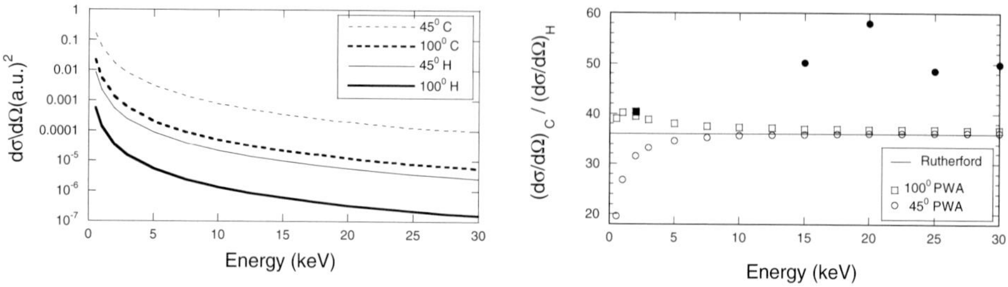

Figure 4. The differential cross section, as calculated in using partial waves [13] as a function of energy at 45° and 100° for carbon and hydrogen atoms (left panel) as well as their ratio (right panel). The calculated cross section ratio (open symbols) approaches with increasing energy the value of $(Z_1/Z_2)^2$ (presented by the full line) as predicted by the Rutherford cross section. Experimental cross section ratios are plotted as well, represented by filled symbols

3. Theory

In the previous section we assumed that the cross section for electron scattering was just proportional to Z^2, and the width was proportional to the momentum component of the proton along $\boldsymbol{k}$. This is a simple picture, relying completely on classical physics. It is just like Rutherford backscattering, using electrons, rather than ions as the probing particles.

Nowadays elastic scattering of electrons from atoms is usually treated as potential scattering. The potential is due to the charge of the nucleus, presented as a point charge, and the charge density associated with the electron wave function. At low momentum transfer the screening of the target electrons is crucial as it cancels the long-range tail of the nuclear Coulomb potential. At high momentum transfer the scattering occurs in a much smaller region around the nucleus and screening should become less important. Within potential scattering theory there is no energy transfer from the electron to the target and thus we can not distinguish electrons scattered from different nuclei. Hence we can not completely describe these measurements by just replacing the spherically symmetric potential of an atom by a molecular potential of lower symmetry.

The scattering at large momentum transfer happens close to a nucleus, where the electronic structure is similar in the molecule compared to the atom. We assume thus that for large momentum transfers scattering from an atom in a molecule, is the same as scattering from a single atom. This can be treated by standard potential scattering theory. If the momentum transfer is so large that in a volume with radius $1/q$ around the nucleus the total electron charge density is small the approximation of the potential by a Coulomb potential should become reasonable. Hence the differential cross section should approach the Rutherford value. In order to investigate this further we calculated the differential elastic cross section for isolated atoms using the partial wave formalism. We used the code written by Salvat and Mayol [13]. Exchange and relativistic effects are included in this code but polarisation effects are not considered. In Fig. 4 we plot the differential cross section as a function of energies for scattering angles of 100 and 45 degrees, as well as the ratio of the carbon and hydrogen cross section. This ratio approaches at higher energy the ratio predicted by the Coulomb cross section within a few percent.

Of course the charge distribution in molecules is modified compared to the free atom.

The main influence on the cross section at high momentum transfer will be due to the carbon 1s electrons, as these electrons are most effective in screening. These electrons are not expected to change significantly due to the chemical bond. Thus we conclude that under the present conditions errors introduced by assuming scattering from a Coulomb field to be of the order of a few percent, an order of magnitude smaller than the observed discrepancies.

In the previous section we use potential scattering theory for obtaining the cross section, and we derive the energy loss of the scattering event from a classical argument based on the finite mass and momentum of the target. For weakly interacting probes such as neutrons we can describe the scattering experiments in the first Born approximation and a single quantum theory has been developed that describes the observed intensity and energy loss distribution. Here we quote some of the results of this theory, a full description can be found in the neutron literature [2, 3, 4]. In the first Born approximation it can be shown that

$$\frac{d^2\sigma}{d\Omega d\varepsilon} = AS(\boldsymbol{q},\omega) \tag{3}$$

$$A = \frac{m^2}{4\pi^2}\frac{k_1}{k_0}W(q)$$

$$W(q) = \left(\int \exp(i\boldsymbol{q}\cdot\boldsymbol{r})V(\boldsymbol{r})d\boldsymbol{r}\right)^2.$$

Here m is the mass of the projectile, $\boldsymbol{k}_0$ and $\boldsymbol{k}_1$ the momentum of the scattered particle before and after the collision, and $V(\boldsymbol{r})$ the particle-target interaction. $S(\boldsymbol{q},\omega)$ is the dynamical structure factor, a target property.

$S(\boldsymbol{q},\omega)$ is the Fourier transform in space and time of the pair distribution function $G(\boldsymbol{r},t)$

$$S(\boldsymbol{q},\omega) = \frac{N}{2\pi}\int e^{i(\boldsymbol{q}\cdot\boldsymbol{r}-\omega t)}G(\boldsymbol{r},t)d\boldsymbol{r}dt \tag{4}$$

with N the particle density [3]. $G(\boldsymbol{r},t)$ is generally a complex function, but is real for classical particles. Then it is simply the density distribution at $\boldsymbol{r}$ and time t as seen from the point $r=0$ where a particle passed at $t=0$. $G(\boldsymbol{r},0)$ is the (real) pair distribution function and is probed in truly elastic scattering (e.g. diffraction).

For large $\boldsymbol{q}$ and $\boldsymbol{r}$ (corresponding to different atom positions) the integrant in Eq. 4 is rapidly fluctuating and will not contribute much to the integral. Thus $G(\boldsymbol{r},t)$ is probed over a region $r \simeq 1/q$ and hence at large q values we have essentially binary collisions with single atoms. (This is in the neutron literature referred to as the incoherent approximation.) The characteristic width of $S(\boldsymbol{q},\omega)$ is proportional to $\overline{p_q}q/m$ with $\overline{p_q}$ the root mean square momentum component of the target atom along $\boldsymbol{q}$ (see Eq. 1). There is no structure in $S(\boldsymbol{q},\omega)$ at a smaller energy scale. That means that the longest characteristic time scale of interest in the Fourier transform is $t = m/(\overline{p_q}q)$ [2]. This time is considered to be the scattering time τ. It decreases with increasing values of q. For the current experimental conditions $\tau \leq 10^{-15}$ sec.

In the previous paragraphs (e.g. in the derivation of Eq. 1) we assume that we can treat the collision between an electron (neutron) with a nucleus as a collision between two free particles. This is usually referred to as the impulse approximation. If the impulse approximation applies $S(\boldsymbol{q},\omega)$ simplifies to [4]:

$$S(\boldsymbol{q},\omega) = \delta(\omega - \frac{q^2}{2m} - \frac{\boldsymbol{p}_0\cdot\boldsymbol{q}}{m}) \tag{5}$$

and hence we can measure a projection of the momentum distribution of the nucleus.

Does this approach apply as well to the present electron scattering measurements? The validity of the Born approximation is not as easily justified for strongly interacting particles such as electrons as it is for weakly interacting neutrons. In the high-energy limit the interaction of an electron with a single atom becomes weak and Eq. 3 should become a reasonable description of the experiment. From measurements of noble gases we know that the differential elastic cross section of $\simeq$ 40 keV electrons from argon is w ell described by the first Born approximation but the approximation breaks down for higher Z species such as xenon [14, 15]. Hence we expect the first Born approximation (and hence the factorisation of Eq. 3) to provide a reasonable description for 25 keV electrons scattering quasi-elastically from molecules composed of light elements, such as carbon, oxygen and hydrogen, but it may fail at lower energies and/or if heavier elements are involved.

The only part of Eq. 3 that depends on the probing particle is the projectile-target interaction $V(\boldsymbol{r})$ and the projectile mass m. As these quantities are known for both neutrons and electrons we can compare neutron and electron scattering experiments to test explicitly if Eq. 3 gives a consistent description of both experiment. A comparative study, using polyethylene as a target has been described extensively in [16] and we summarise the main points here.

In order to have similar momentum transfer as in our electron scattering experiment one has to scatter neutrons with energies around 20 eV. These NCS experiments were done at the Vesuvio beam line of ISIS and use a pulsed neutron beam in combination with filters with sharp absorption dips to determine the neutron energy loss by a time-of-flight technique [17]. Let us first compare the proton momentum distribution as measured by electron and neutron scattering. This is displayed in the left panel of Fig. 5. Using Eq. 1 or Eq. 5 we can calculate from the measured energy transfer distribution the proton momentum distribution. In the left panel of Fig. 5 we plot the width of the proton momentum distribution as a function of momentum transfer. The proton momentum distribution is indeed independent of the momentum transfer, and within the spread of the measurement, the same results are obtained for protons and neutrons. In the right panel of Fig. 5 we plot the R_{obs}/R_{conv} values found for both technique as a function of momentum transfer. For neutrons one can calculate the same ratio R_{obs}/R_{conv} based on the known neutron cross sections of H and C [18]. For both electrons and neutrons R_{obs}/R_{conv} is systematically less than 1 by a similar amount. For neutron scattering these effects are known from a large group of proton containing materials [19, 20, 21]. Thus also the anomalous neutron scattering cross section seems to be reproduced by the electron experiment. Similar agreement between the electron and neutron measurement was found for formvar, a different polymer [22].

In summary at high-enough incoming energy and momentum transfer we expect the results of electron scattering to be directly comparable with the neutron data. These conditions are probably well full-filled at the high energies described here (30 keV) but more questionable for the 2 keV experiment. Indeed in the comparative study of neutron and electron scattering experiments of polyethylene good agreement was found between both techniques.

In reality the nucleus is bound as it is part of a molecule. The energy transfer to the proton is of the order of the bond energy. However the impulse approximation can still be valid, even if the energy transfer is less than the bond strength [2]. One way of explaining the working of the impulse approximation under these conditions is to say that during the interaction time the nucleus moves over such a small distance that it does not feel any change in potential. The interaction time is inversely proportional to the transferred momentum. It does not relate to the 'velocity' of the scattered particles. Note that the velocity of a 25 keV

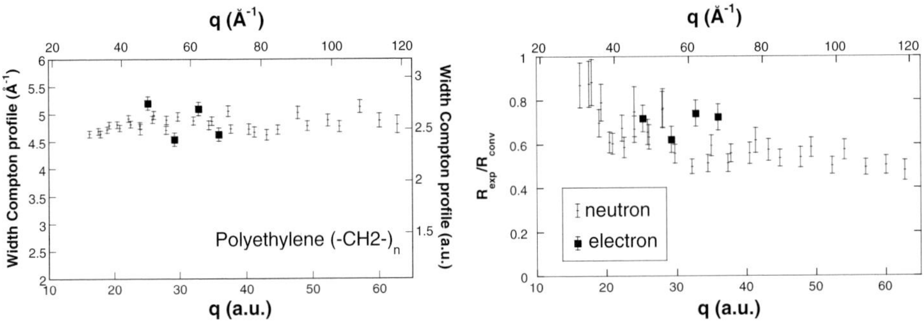

Figure 5. Results for a polyethylene film as obtained by neutrons and electrons. A comparison of the dependence of the width of the Compton profile on the momentum transfer is shown in the left panel. The observed hydrogen-to-carbon peak area ratio R_{exp} divided by the calculated ratio R_{conv}, based on the calculated cross sections, is shown in the right panel.

electron is more than 1000 times that of the velocity of a 25 eV neutron. Thus the proposition that the impulse approximation is independent of the velocity of the probing particle could be tested in a suitably-designed experiment using both electrons and neutrons.

$G(\boldsymbol{r},t)$ is a function of $\boldsymbol{r}(t)$: the Heisenberg operator $exp(itH)r(0)exp(-itH)$. Hence for $t > 0$, $G(\boldsymbol{r},t)$ becomes complex [3]. The protons and electron wave function are coupled by the Coulomb matrix elements and the system will not separate in the product of an electron wave function and a proton wave function. For large t the proton will have interacted with the environment and the phase information is lost. Due to this decoherence the proton will have lost the capability to interfere, and acts like a classical particle, i.e. $G(\boldsymbol{r},t)$ becomes real again. For neutron scattering this theory is formulated in terms of density matrices [23], and proposes that if the interaction time of the neutron with the target is of similar order of magnitude as the decoherence time, a reduced differential cross section should be observed in neutron scattering. Similar considerations would apply to electron scattering. Thus if the momentum transfer is high enough the scattering time will become comparable to the decoherence time, and the measurement should probe the proton wave function in the entangled proton-electron system.

In the impulse approximation the transfer of momentum to the proton does not affect the electronic structure. The bond may break, leading to electronic excitations, but this happens on a slower time-scale than the collision itself. Note that the bond strength of a C-H bond is $\simeq 4.5$ eV, larger than the energy transfer in the case of CH_4, but smaller than the energy transfer in the high-energy PE experiment. For the correlated electron-proton system the usual assumption that momentum can be transferred to the proton, without a probability of direct electronic excitations, could turn out to be wrong. This could transfer intensity from the proton peak to the background and hence be the cause of the missing proton intensity.

Electron-electron correlation gives rise to satellites in (e,2e) measurements, reducing the intensity of the main peak (see e.g. McCarthy *et al* [24]). Electron-proton correlation could have similar effects on the strength of the proton-derived elastic scattering peak.

4. Conclusions and Outlook

We measured the energy distribution of electrons scattered quasi-elastically from hydrocarbons both in the form of films and in the gas-phase. These experiments reveal information on the momentum distribution of the protons. There are currently problems reconciling the signal strength of the proton derived signal compared to the carbon derived signal for the solid films. Similar problems are present in neutron scattering experiments operating at comparable momentum transfer. Although there is currently not a full-fledged theory making quantitative predictions it is suggested that these discrepancies could be due to the quantum nature of the protons, probed at the short time scale of the collisions.

Especially the gas-phase measurement could provide a fertile testing ground of theory. Here there are no problems with energy-dependent backgrounds and/or radiation damage. Also neutron scattering can not as easily measure these low-density targets. Extending these measurements over a larger energy range could test the underlying theory (partial wave analysis, polarisation effects, impulse approximation and the possible quantum effects) in a way not easily accomplished by other means. The possibility of transferring kinetic energy to a single atom within a molecule is a fascinating one, especially if this energy is of the order of the chemical bond strength. Unfortunately the cross sections seem to be too small to detect molecular fragments in coincidence with the energy loss spectra, which would make the study of the break-up of molecules under these conditions possible.

References

[1] H. Boersch, R. Wolter, and H. Schoenebeck, Z. Physik **199**, 124 (1967).

[2] G. I. Watson, J. Phys.: Condens. Matter **8**, 5955 (1996).

[3] L. V. Hove, Phys. Rev. **95**, 249 (1954).

[4] V. F. Sears, Phys. Rev. B **30**, 44 (1984).

[5] M. Vos, G. P. Cornish, and E. Weigold, Rev. Sci. Instrum. **71**, 3831 (2000).

[6] B. Cranfill, Rev. Sci. Instrum. **49**, 264 (1978).

[7] Y. K. Godovsky and S. N. Magonov, Langmuir **16**, 3549 (2000).

[8] M. Vos, Phys. Rev A **65**, 12703 (2002).

[9] M. P. Paoli and R. S. Holt, J. Phys. C: Solid State Phys. **21**, 3633 (1988).

[10] J. C. Corchado and D. G. Truhlar, J. phys. Chem. A **102**, 1895 (1998).

[11] J. Mayers, T. M. Burke, and R. J. Newport, J. Phys.: Condens. Matter **6**, 641 (1994).

[12] A. Hitchcock, J. Electron Spectrosc. Relat. Phenom. **112**, 9 (2000).

[13] F. Salvat and R. Mayol, Comp. Phys. Commun. **74**, 358 (1993).

[14] M. Fink and J. Kessler, Z. Physik **196**, 1 (1966).

[15] M. Fink and R. A. Bonham, Phys. Rev. **187**, 114 (1969).

[16] M. Vos, C. Chatzidimitriou-Dreismann, T. Abdul-Redah, and J. Mayers, Nucl. Instr. and Meth. B in press (2004).

[17] R. M. Brugger, A. D. Taylor, C. E. Olsen, J. A. Goldstone, and A. K. Soper, Nucl. Instrum. and Meth. **221**, 393 (1984).

[18] S. Lovesey, *Theory of Neutron Scattering from Condensed Matter* (Clarendon Press, Oxford, 1984), Vol. 1.

[19] C. A. Chatzidimitriou-Dreismann, T. Abdul-Redah, R. M. F. Streffer, and J. Mayers, Phys. Rev. Lett. **79**, 2839 (1997).

[20] C. A. Chatzidimitriou-Dreismann, T. Abdul-Redah, and R. M. Streffer, J. Chem. Phys. **116**, 1511 (2002).

[21] E. B. Karlsson, T. Abdul-Redah, R. M. F. Streffer, B. Hjörvarsson, J. Mayers, and C. A. Chatzidimitriou-Dreismann, Phys. Rev. B **67**, 184108 (2003).

[22] C. A. Chatzidimitriou-Dreismann, M. Vos, C. Kleiner, and T. Abdul-Redah, Phys. Rev. Lett. **91**, 57403 (2003).

[23] C. A. Chatzidimitriou-Dreismann, J. Alloys Comp. **356-357**, 244 (2003).

[24] I. E. McCarthy, R. Pascual, P. Storer, and E. Weigold, Phys. Rev. A **40**, 3041 (1989).

Inst. Phys. Conf. Ser. No. 183
Paper presented at Int. Conf. Electron and Photon Impact Ionization and Related Topics, Louvain-la-Neuve, 2004

Multiple ionization in strong fields

A. Rudenko, R. Moshammer, C.D. Schröter, K. Zrost, B. Feuerstein, V.L.B. de Jesus, J. Ullrich
Max-Planck-Institut für Kernphysik, Heidelberg, Germany

Abstract. Single and multiple ionization of rare gas atoms by 25 and 7 fs intense laser pulses has been studied using a "Reaction Microscope". For single ionization of He, Ne and Ar in the tunnelling regime, surprising and distinct patterns have been observed in high-resolution low-energy electron spectra. Mechanisms of double and multiple ionization were elucidated and partly identified via recoil-ion momentum distributions and correlated electron spectra. For Ne the highly correlated "instantaneous" emission of up to four electrons is triggered by a recollisional electron impact, whereas in multiple ionization of Ar different mechanisms involving field ionization steps and recollision-induced excitations play a major role.

1. Introduction

The interaction of intense coherent light with atoms, molecules, clusters or solids has attracted increasing attention in the recent past, mainly driven by the rapid development of laser technology (see e.g. [1]). Femtosecond pulse-lengths down to two optical cycles in the visible are at hand [2] and intensities of up to 10^{19} W/cm^2 have been reached [3] using high repetition-rate, "table-top" Ti:Sa lasers (800 nm, 1.5 eV). Attosecond pulses were demonstrated in high-harmonic radiation [4], absolute phases between the carrier-wave and the envelope have become controllable [5], attosecond pump-probe techniques are realized [6], phase and polarization shaping is applied [7] and VUV to x-ray free-electron laser sources are on the horizon [8]. This tremendous technological progress opens new perspectives for experiments and theory. Many features of strong-field induced phenomena, such as single electron emission, the intensity dependence of ion rates, multiple ionization yields, dissociation of molecules, harmonic generation etc. have been explored in detail over many years essentially using single particle detection techniques (see e.g. [9]). Thus, a profound theoretical understanding of non-linear multi-photon processes has emerged for mechanisms that either involve only one active electron or where electrons can be considered to act independently within a single active electron (SAE) description (see e.g. [10]).
Still however, even the very basic process, atomic single ionization, has resisted consistent and complete theoretical treatment at high laser intensities, where the "multi-photon" regime enters into the strong-field limit and field- or tunnel-ionization is usually expected to dominate [11]. Moreover, the prototype two-electron reaction, helium double ionization, has remained the subject of numerous controversial debates over the last decade [12]. The *ab-initio* description of true many-electron dynamics by explicitly solving the many-particle time-dependent Schrödinger equation, seems to be far beyond present computing capabilities for systems larger than helium, the H_2^+ molecule or the H^- negative ion.

The recent introduction of next-generation, high-resolution many-particle imaging technologies ("Reaction Microscopes": [13,14]) into the field [15,16], now allows one to explore the correlated motion of several electrons and ions, their time-dependent dynamics in the field, pathways to multiple ionization, the dissociation of molecules or even single electron emission with increasing sophistication and accuracy, challenging theory (see also [17]). As it turns out the detection of final-state momenta provides access to absolute phases and time-delays where fragmentation or particle emission occurs in the oscillating laser field with an effective sub-laser cycle time-resolution. Thus, in many cases mechanisms of single and multiple ionization can be identified and discriminated.

The present paper summarizes some new experimental results on the fragmentation of atoms obtained using these next-generation experimental methods. After a brief outline of the imaging technique in section 2, a selection of illustrative experimental results on single and multiple ionization with 25 and 7 fs laser pulses is presented in section 3. Conclusions and future perspectives are envisaged in section 4.

2. Experiment: Reaction Microscope

Reaction Microscopes – "the bubble chambers of atomic and molecular physics" – enable to determine the momentum vectors of several electrons and ions being ejected as the result of the fragmentation of atoms, molecules or clusters in any kind of external fields [13,14,18]. Since four years, the technique has been successfully applied to explore in unprecedented detail pathways to single and multiple electron ejection in femtosecond PW/cm^2 laser fields. Experiments with intense lasers are exceptionally challenging for the following reasons: First, the residual gas pressure as well as the target density has to be extremely small, since the PW/cm^2 laser pulse ionizes essentially all particles within its focus. Thus, as correlated emission of electrons and ions from the same atom shall be measured, only one target atom is allowed to be in the focus of the laser pulse and, accordingly, the target is operated at a very low density at a background pressure of $2 \cdot 10^{-11}$ Torr. Second, since no more than one atom should be ionized per laser shot in order to maintain information about the correlation, high repetition rates are mandatory. Whereas charged particle or single photon impact experiments run with a typical repetition rate of up to 1 MHz, state-of-the-art high-power Ti:Sa lasers operate at kHz repetition rates. For this reason, though typical measuring times are several days, all laser experiments involving coincidence measurements still suffer from limited statistical significance.

A detailed description of a state-of-the-art Reaction Microscope dedicated for the laser experiments can be found in [17]; here we briefly summarize the salient features of the setup. As illustrated in Figure 1, the laser beam is focused by a spherical on-axis mirror to a typical diameter of 5-10 μm onto a low-density (10^8 -10^9 atoms/cm^3) supersonic, internally cold atomic gas-jet target. Recoiling ions and electrons emitted during the reaction are guided to two multi-hit, position-sensitive channel plate detectors by homogeneous electric and magnetic fields applied perpendicular to both laser and atomic beam propagation directions. From the times-of-flight and the positions of arrival, the initial vector momenta of all the reaction fragments are calculated from the equations of motion for electrons and ions in the well-known electric and magnetic fields.

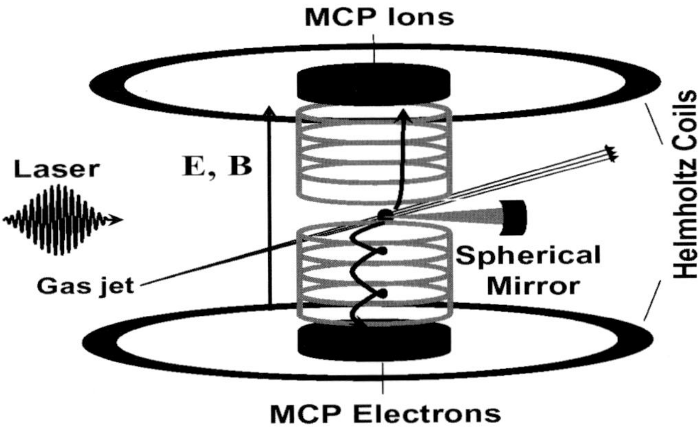

Figure 1. Experimental set-up.

By varying the magnitude of the projection fields, both the resolution as well as the acceptance for the fragments in momentum space that are projected, can be chosen over a wide range. Typically, for one setting, all ions of interest with momenta $|P_R| < 5$ a.u. are accepted simultaneously (a.u.: atomic unit). At the same time, all electrons with transverse (perpendicular to the extraction field) energies $E_{e\perp} < 50$ eV as well as with longitudinal (parallel to the extraction field) energies of $E_{e\|} < \infty$ eV into the forward (towards the detector), and $E_{e\|} < 15$ eV into the backward directions are detected. Up to ten hits on the electron detector can be accepted for a minimum time between two hits of 10 ns. The position resolution of the delay-line read-out system can be as good as 100 μm. Depending on the magnitude of electric (~ 1-2 V/cm) and magnetic fields (typically 5-20 G) chosen in the specific experiment the Reaction Microscope is able to simultaneously monitor up to 90 % of the nine-dimensional final-state momentum space for double ionization.

The ion momentum resolution is limited by the internal jet temperature and is about 0.2 a.u. along the jet propagation direction. However, by using improved jet geometry and by collimating the atomic beam, a superior resolution of $\Delta P_\| \sim 0.02$ a.u. is achieved in the direction of the extraction field (perpendicular to the jet). The electron momentum resolution is $\Delta P_{e\|} \sim 0.02$ a.u. along the extraction, while the resolution in the transverse direction depends on the time-of-flight. It varies from zero for flight times corresponding to any integer multiple of the cyclotron revolution period, where electrons are focused to the center of the image independent on their momenta in the detector plane, to ~ 0.05 a.u. for the electrons arriving exactly in the middle of the period.

For the experiments described below, linearly polarized radiation of a Kerr-lens mode locked Ti:sapphire laser at 795 nm wavelength amplified to pulse energies of up to 350 μJ at 3 kHz was used. The polarization direction was oriented parallel to the extraction field. The width (FWHM) of the amplified pulses was 25 fs. In order to generate few-cycle pulses they are spectrally broadened in a gas-filled hollow fiber and then compressed to 6-7 fs (FWHM) by chirped mirrors and a prism compressor. Fluctuations of the laser intensity from pulse to pulse were monitored during the experiment and did not exceed 5%. Absolute calibration of the peak intensity was performed using a clear kink in the measured momentum distribution of the electrons from single ionization of Ne, which corresponds to the maximum drift momentum of $2\sqrt{U_p}$ that electrons can gain from the laser field [19].

3. Results

3.1 Single ionization: resonant-like features of the low-energy electron continuum

When an atom is exposed to an intense laser field, ionization is usually considered either as a multiphoton or tunnelling process depending on the value of the so-called Keldysh, or adiabaticity parameter $\gamma = \sqrt{Ip/2Up}$ (where Ip is the ionization- and $Up = I/4\omega^2$ the po n-deromotive potential, respectively. I is the light intensity and ω its frequency) [20]. For $\gamma >$ 1 numerous experiments revealed the multiphoton nature of ionization, reflected in a rich structure in the photoelectron spectra. Various field induced processes, such as resonant and non-resonant above-threshold ionization (ATI), ponderomotive effects, channel-opening and closing etc. were studied, and good agreement between theory and experiment has been achieved (see e.g. [21]).

For the opposite case of very low γ where tunnel ionization is assumed to dominate, only few experimental results have been reported. They show rather smooth and unstructured photoelectron spectra in general agreement with the quasistatic tunnelling description, which assumes that the electron first tunnels into the continuum and then moves classically gaining drift energy from the oscillating laser field [22]. However, deviations from this relatively simple semiclassical picture were found first for the high-energy region of the photoelectron spectra (see, e.g. [23]), and, very recently, for ultra low-energy electrons [11]. Most of these observed features were explained to be due to the rescattering of the emitted electron on its parent ion. However, the details of the recollision process, the role of the atomic structure and interference effects are still subjects of discussions (see e.g. [24] and references in [11]), and no consistent and complete theoretical description of single electron emission has yet emerged for laser intensities that might be characterised as the transition between the multiphoton and tunnelling regime ($\gamma \leq 1$).

Figure 2 displays momentum distributions of singly charged ions along the laser polarisation axes for three different targets at various intensities (γ varies from 1.1 to 0.29) for 25 fs (2a-c) and 7 fs (2d) laser pulses. The data are integrated over all transverse momenta. For 25 fs and all targets the spectra at the lowest intensity manifest a clear set of peaks that are broadened as the intensity grows. However, for all intensities there is a qualitative difference between Ar and the two other species: while He and Ne exhibit a minimum at zero longitudinal momentum, there is a clear maximum for Ar. Comparing results obtained with 25 fs and 6-7 fs pulses for Ne (Fig. 2c and 2d), one can see that for the shorter pulse duration the peaks are smoothened, and, with the exception of the minimum at zero, an otherwise unstructured distribution is observed at high intensities.

Figure 3a shows electron energy distributions for Ne with 25 fs pulses. Again, the spectra consist of a series of peaks separated by the photon energy and broadened with increasing intensity. This is the first observation of ATI-like structures in the photoelectron energy spectra at laser intensities deep in the tunneling regime (1 PW/cm^2 and higher). In both, the momentum and energy representations, the peaks do not exhibit intensity-dependent shifts (see vertical dashed lines in Figures 2-3). Figure 3b shows in a larger scale the longitudinal part of the electron kinetic energy (integrated over the transverse electron energy). Here some fine structure with two sub-peaks can be seen, repeated within different ATI orders. These substructures are not resolved in the total kinetic energy spectra, most likely due to the fact that the momentum resolution in the transverse direction is not as good as in the longitudinal direction, resulting in a worse total energy resolution.

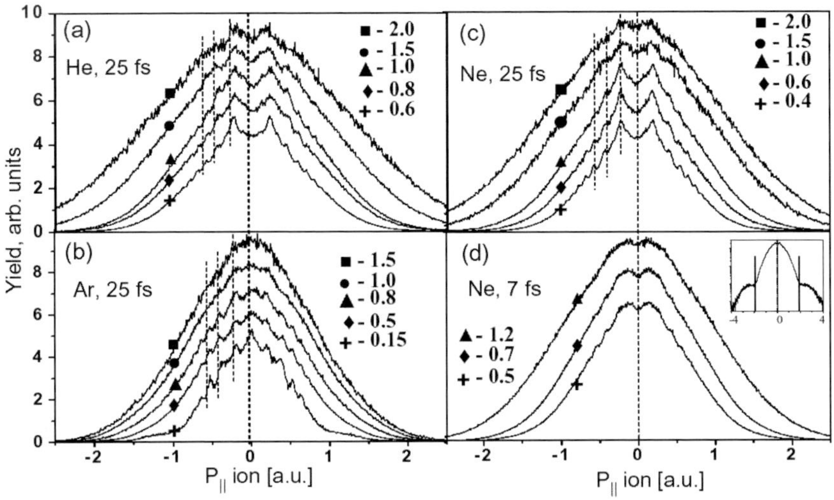

Figure 2. Momentum distributions of singly charged ions along the laser polarization direction. Intensities in PW/cm^2 are indicated in the figure. The inset shows in logarithmic scale the spectrum used for the absolute intensity calibration, where the vertical lines indicate the momentum $\pm 2\sqrt{Up}$ (see text).

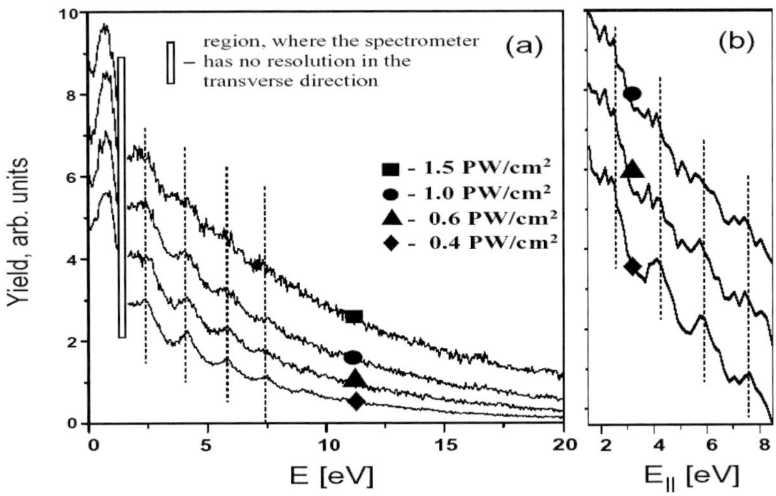

Figure 3. Electron energy distributions for single ionization of Ne by 25 fs laser pulses. (a): total kinetic energy. (b): longitudinal part of the kinetic energy.

The splitting of ATI peaks and the absence of any detectable pondermotive shifts are the typical features of the resonant-enhanced ionization. It is nowadays generally accepted that the multiphoton peaks from nonresonant ionization shift towards lower energies as the intensity grows, due to the field-induced increase of the ionization potential, which is approximately equal to the ponderomotive energy Up. Thus, for values of Up much larger than the photon energy, the averaging over different intensities due to the spatio-temporal structure of the pulse or just a result of intensity fluctuations would smear out structures originating from nonresonant ionization. On the other hand, multiphoton resonances with certain excited atomic level always occur at the same intensity (same Up!). If the peak intensity is higher than the resonant value, the resonance condition for one or several excited levels can always be fulfilled somewhere within the laser focus, producing the intensity independent fine structure in the ATI spectra.

The assumption that resonances lead to the observed structures can explain the atomic species dependence observed around zero momentum as well as the degeneration of the peaks for 6-7 fs pulses because contributions from any resonant-like mechanism should be sensitive to the duration of the laser pulse [25]. However, the direct applicability of this consideration for the case of high intensities, where all excited states are strongly deformed by the laser field, is somewhat questionable. An alternative approach traces the origin of the resonant-like structures in the spectra to the so-called "channel-closing", multiphoton resonances with the ground state [26].

3.2 Multiple ionization: correlated few-electron dynamics

While single electron emission can be essentially considered within SAE approach, the latter fails severely for the case of double and multiple ionization. It is known since more than 20 years [27] that the yields of doubly and multiply charged ions produced by linearly polarized laser pulses exceed significantly those expected for independent successive removal of two or more electrons over wide intensity regimes. The origin of this enhancement, which was called "non-sequential" ionisation, remained controversial until 2000 though there was a general agreement that electron-electron correlation should be involved. First measurements of recoil-ion momentum distributions [15,16] along with the results of earlier experiments using circularly polarized light [28,29] provided convincing evidences in favour of the so-called "recollision" model [30]. Since then, for double ionization ion momentum distributions, coincident photoelectron energy spectra and correlated electron momenta were measured for different rare gases in a broad range of experimental parameters. However, the details of the double ionization attosecond dynamics are still far from being completely understood. In particular, one of the most intriguing questions is whether and how atomic structure can influence ionization dynamics [19].

For the case when more than two electrons are ionized, there are much less experimental results available. Intensity dependences of ion yields were measured for different rare gases (see, e.g. [31]). The ion momentum distribution obtained in [16] for Ne^{3+} at 1.5 PW/cm^2 remains, to the best of our knowledge, the only differential data-set published up to now. Theoretically, the description of multiple ionization represents an even greater challenge compared to double ionization, not only because of increased number of reaction partners, but also due to the larger variety of the pathways leading to the same final state, including different combinations of sequential and non-sequential processes. As it was shown in [32] for Ar, neglecting specific non-sequential channels leads to a large discrepancy between measured and calculated yields at the intensities ~ 10^{14} - 10^{15} W/cm^2.

Fig. 4a displays intensity dependences of the ratios of the ionization yields for Ne^{2+}/Ne^{+} and Ne^{3+}/Ne^{+}. Fig. 4b shows the same ratios for Ar. Ar^{2+}/Ar^{+}, Ar^{3+}/Ar^{+} and Ne^{2+}/Ne^{+} ratios exhibit a characteristic change in slope that resembles the 'knee' structure observed in the total yields indicating the transition from non-sequential to sequential production of the corresponding charge state. For the production of Ne^{3+} ions the intensity was not high enough to reach the sequential regime.

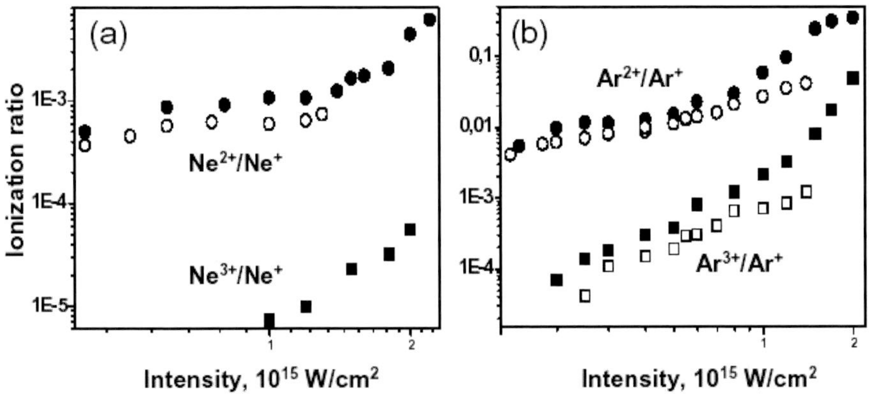

Figure 4. The ratios of double to single and triple to single ionization yields for Ne (a) and Ar (b) targets. Full symbols: 25 fs; open symbols: 7 fs

The results presented in Fig. 4 clearly show, that the relative yields of different charge states depend on a pulse length. For the case of Ar we can cover the transition from the non-sequential to the sequential regime for both, 25 fs and 7 fs pulses. The difference between Ar^{2+}/Ar^{+} (Ar^{3+}/Ar^{+}) ratios for different pulse durations becomes more pronounced in the sequential regime, where the probability of the sequential process obviously depends on the pulse length. Nevertheless, similar to the results of [33], we observe weaker production of doubly charged ions for a shorter pulse also in the non-sequential regime. In [33] this difference was attributed to the Coulomb focusing of electron trajectories on the parent ion after several laser cycles, which is less efficient for a few-cycle pulse. Another reason for this reduction might be a smaller contribution from recollision-induced excitation with subsequent field ionization (RESI): if the probability for the excited state to be ionized within one optical half-cycle is less than unity, the total yield of doubly charged ions produced by this mechanism will depend on the duration of the remaining part of the pulse.

Figure 5 shows momentum distributions of $Ne^{2\text{-}4+}$ ions along the laser polarization direction. All spectra exhibit a clear double peak structure which is a signature of the recollision process. In the momentum distributions of Ne^{3+} and Ne^{4+} ions this structure is even more pronounced than for double ionization, with almost no ions produced with zero momentum. The momentum distribution of Ne^{2+} ions created by a 7 fs laser pulse (solid line in Fig. 5a) closely follows the results obtained with 25 fs pulses, indicating that the momentum distribution is mainly defined by the recollision within one or two cycles after the

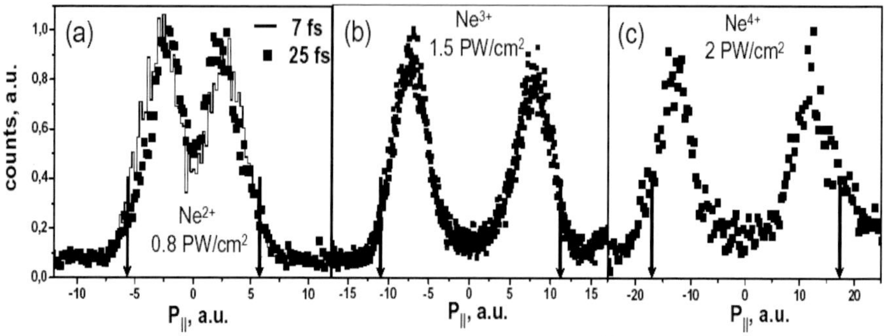

Figure 5. Longitudinal momentum distributions of Ne^{n+} ions. Arrows indicate ion momenta of $\pm 2n\sqrt{U_P}$.

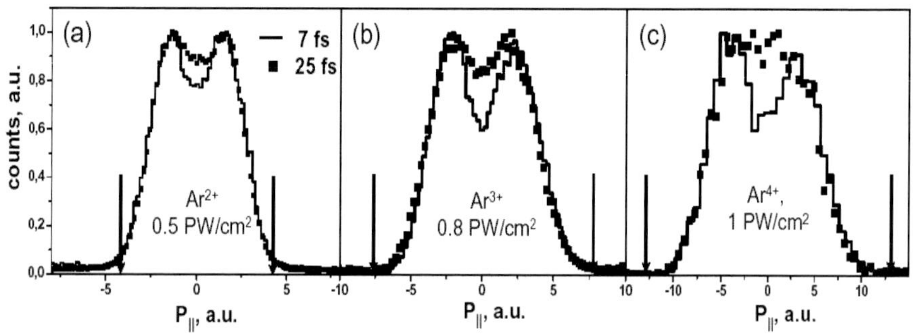

Figure 6. Same as Figure 5 for Ar.

removal of the first electron. The widths of the spectra are in good agreement with kinematical constraints following from classical considerations [34]. In general, confirming the results of [16], we can state that ion momentum distributions of Ne^{n+} ions agree well with a model assuming direct impact ionization by the rescattered electron via an (e,ne) process.

Inspecting the momentum distributions of $Ar^{2\text{-}4+}$ ions (Fig. 6), we must conclude that other mechanisms play an important role. The minima at zero momentum along the longitudinal direction for $Ar^{2\text{-}3+}$ ions obtained with 25 fs pulses (squares in Fig. 6a,b) are much less pronounced than those for Ne. For Ar^{4+} ions created by a 25 fs laser pulse we do not observe a double peak structure at any intensity! This ‘filling of the valley’ can be attributed (at least for the case of double ionization) to the contribution of electron-impact excitation during recollision followed by field ionization of the excited ion in one of the subsequent cycles of the oscillating laser field (RESI). Here, the second ionization step as well occurs close to the maximum of the field, leading to relatively small ion momenta. Theoretical analysis of collision-induced ionization and excitation cross sections shows that direct electron impact ionization clearly dominates for Ne, whereas for Ar and He the amplitude involving excitation plays a decisive role [19]. Signatures of these two mechanisms have also be observed in different emission patterns of two correlated electrons [17,35-37].

As it was pointed out above, the contribution of the RESI mechanism might be suppressed for a shorter laser pulse. Comparing the Ar results obtained for two different pulse lengths, we clearly see that the fraction of ions with zero momentum is considerably reduced for a 7 fs pulse. Here, the minimum at zero momentum is more pronounced for double and triple ionization, and can be also observed for Ar^{4+} ions. The same depletion of the region around zero momentum could also be caused by the suppression of sequential ionization from the ground state. However, the contribution of purely sequential production of Ar^{2+}, $Ar^{3+}$and Ar^{4+} at corresponding intensities is still a small fraction of the total yield [32]. Nevertheless, for the production of triply and fourfold charged ions there is a possibility to reach the final state via a combination of sequential and non-sequential processes, and this dynamics might be changed for a few-cycle pulse. Contributions of such combined processes can also explain the relative narrowing of Ar^{3+} and Ar^{4+} momentum distributions (compare the position of the arrows at Fig. 5 and 6) [34].

4. Conclusions and outlook

We have studied the evolution of ATI-like structures in electron spectra over a broad range of Keldysh parameters ($1.1 > \gamma > 0.29$). We have shown that low-order ATI peaks with some target-dependent substructure can be resolved deep in the tunnelling regime. The peaks do not exhibit ponderomotive shifts and are less pronounced for ionization by few-cycle laser pulses. These are typical features for resonant-enhanced ionization, well-known for the case of much lower intensities. The importance of resonant enhancement and its connection to recollision phenomena was recently demonstrated for the plateau in the photoelectron spectra [38]. Taking into account latest results showing that recollision effects also manifest themselves at low electron energies [11,39], we conclude that the structures presented in our data originate from a similar type of resonant process. However, a detailed description of the intermediate states involved, their relation to the field-free atomic levels and their influence on the electron energy and angular distributions remains the subject of forthcoming analysis.

We found that essentially only one pathway based on the "instantaneous" interaction between up to four electrons dominates multiple electron removal from Ne. Surprisingly, triply and fourfold charged ions are predominately produced by highly correlated sub-cycle few-electron emission even at those intensities where the dominating mechanism of double ionization does not involve electron correlation. For Ar, instead, different types of less correlated multi-electron pathways involving the RESI mechanism and cascades of sequential and non-sequential processes are important. Especially interesting is the fact that the atomic structure plays an important role even at intensities well above 10^{15} W/cm^2 by efficiently 'selecting' the most effective dynamical pathway to multiple ionization, i.e. the coupling between several atomic electrons and the laser field.

The demonstrated possibility to influence many-electron dynamics by changing the laser pulse parameters can allow for a control over various field-induced phenomena. Presently, work is in progress to implement carrier-envelope phase stabilization, to realize medium-power MHz oscillators in order to increase the pulse repetition rate and to produce ultra-short pump-probe pulses for the manipulation and possible control of electronic wavepackets on very short time scales. Exciting possibilities to explore new non-linear multiphoton phenomena, where only a few high-energy photons interact with few electrons, will emerge with the beginning of the experimental program at the novel free-electron laser, the TESLA Test Facility in Hamburg in 2005 and experiments are under preparation.

References

[1] Brabec T and Krausz F 2000 Rev. Mod. Phys. **72** 545
[2] Nisoli M et al. 1997 Opt. Lett. **22** 522
[3] Walker B et al. 1999 Opt. Exp. **5** 196
[4] Paul M P et al 2001 Science **292** 1699
[5] Paulus G G et al. 2003 Phys. Rev. Lett. **91** 253004
[6] Kienberger R et al. 2004 **423** 817
[7] Suzuki T, Minemoto S, Kanai T and Sakai H 2004 Phys. Rev. Lett. **92** 133005
[8] Ayvazyan V et al. 2002 Phys. Rev. Lett. **88** 104802
[9] DiMauro L and Agostini P 1995 Advances in Atomic and Molecular Physics (Academic Press), New York
[10] Protopapas M, Keitel C and Knight P 1997 Rep. Prog. Phys. **60** 389
[11] Moshammer R et al. 2003 Phys. Rev. Lett. **91** 113002
[12] Dörner R et al. 2002 Adv. in Atom., Mol., and Opt. Physics **48** 1
[13] Dörner R et al. 2000 Phys. Rep. **330** 95
[14] Ullrich J et al. 2003 Rep. Prog. Phys. **66** 1463
[15] Weber Th et al 2000 Phys. Rev. Lett. **84** 443
[16] Moshammer R et al. 2000 Phys. Rev. Lett. **84** 447
[17] De Jesus V L B et al. 2004 J. Electron Spectrosc. Relat. Phenom., in print
[18] Moshammer R et al. 1996 Nucl. Instr. Meth. B **107** 62
[19] De Jesus V L B et al. 2004 J. Phys. B **37** L161
[20] Keldysh L V 1965 Sov. Phys. JETP **20** 1307
[21] Wiehle R, Witzel B, Helm H and Cormier E 2003 Phys. Rev. A **67,** 063405
[22] Delone N B and Krainov V P 1998 Phys. Usp. **41** 469
[23] Walker B, Sheehy B, Kulander K C and DiMauro L F 1996 Phys. Rev. Lett. **77** 5031
[24] Wassaf J, Véniard V, Taïeb R and Maquet A 2003 Phys. Rev. A **67**, 053405
[25] Grasbon F et al. 2003 Phys. Rev. Lett. **91** 173003
[26] Kopold R, Becker W, Kleber M and Paulus G G 2002 J. Phys. B **35** 217
[27] L'Huiller A, Lompre L, Mainfray G and Manus C 1983 Phys. Rev. A **27** 2503
[28] Fittinghoff D N, Bolton P R, Chang B and Kulander K C 1994 Phys. Rev. A **49** 2174
[29] Dietrich P et al. 1994 Phys. Rev. A **50** R3585
[30] Corkum P B 1993 Phys. Rev. Lett. **71** 1994
[31] Larochelle S, Talebpour A and Chin S L 1998 J. Phys. B **31**, 1201
[32] Becker A and Faisal F HM 1999 J.Phys.B **32**, L335
[33] Bhardwaj V R et. al. 2001 Phys. Rev. Lett. **86**, 3522
[34] Feuerstein B, Moshammer R and Ullrich J 2000 J. Phys. B **33** L823
[35] Weber Th et al 2000 Nature **405** 658
[36] Feuerstein B et al. 2001 Phys. Rev. Lett. **87** 043003
[37] Moshammer R et al. 2003 J. Phys. B **36** L113
[38] Nandoor M J, Walker M A, van Woerkom L D and Muller H G 1999 Phys. Rev. A, **60** R1771
[39] Chen J and Nam C H 2002 Phys. Rev. A **66** 053415

Inst. Phys. Conf. Ser. No. 183
Paper presented at Int. Conf. Electron and Photon Impact Ionization and Related Topics, Louvain-la-Neuve, 2004

Complete experiments in photoionization with excitation: Theory and Practice

A N Grum-Grzhimailo†and M Meyer‡

†Institute of Nuclear Physics, Moscow State University, Moscow 119992, Russia

‡L.U.R.E., Centre Universitaire Paris-Sud, Bâtiment 209D, F-91898 Orsay Cedex, France

Abstract. An overview on complete experiments in atomic photoionization with excitation is given. Photoionization with the final ion in a discrete excited state is discussed in detail. Polarimetry of the photoion fluorescence provides relative partial photoionization cross sections and can complete the photoelectron spectrometry data to obtain the photoionization amplitudes. Photoionization with excitation from the outer shell of Xe is considered as an example.

1. Definition of complete experiment

Quantum reactions are completely characterized by the absolute magnitudes and phase differences of the complex amplitudes of transitions between the initial state and each of the degenerate final states (channels). A complete set of experimental data is one from which all the amplitudes can be extracted within a certain theoretical model [1]. From this information any observable can be predicted and the most comprehensive test of theory can be provided up to the degree allowed by quantum mechanics.

Let a state of a system be characterized before the reaction by the density operator ρ_0. After the reaction the system is transformed into a state described by the density operator $\rho = S\rho_0 S^+$, where S is the S-operator in the scattering theory. Assume that the detector(s) of the final state is 'tuned' to a state described by the density operator ε, which means that the registration occurs only, if the detector is affected by the system in the state ε. With appropriate normalization of the density operators, the intensity of the detected signal is $W = \mathrm{Tr}\,\rho\varepsilon$. Choosing convenient basis sets, $|\xi_i\rangle$ and $|\chi_f\rangle$, for the initial and final states of the system, respectively, we obtain

$$W = \sum_{ff'} \langle \chi_f | T\rho_0 T^+ | \chi_{f'} \rangle \langle \chi_{f'} | \varepsilon | \chi_f \rangle = \sum_{ff'ii'} \langle \xi_i | \rho_0 | \xi_{i'} \rangle \langle \chi_{f'} | \varepsilon | \chi_f \rangle T_{i\to f} T^*_{i'\to f'} \quad (1)$$

where $T = S - I$ is the transition operator, I is the unity operator. We have introduced the transition amplitudes $T_{i\to f} \equiv \langle \chi_f | T | \xi_i \rangle$. The coefficients in the bilinear combination (1) of the amplitudes $T_{i\to f}$ are known within a certain theoretical framework. Tuning the detector to another state, ε', another bilinear combination of the amplitudes can be measured

$$W' = \sum_{ff'ii'} \langle \xi_i | \rho_0 | \xi_{i'} \rangle \langle \chi_{f'} | \varepsilon' | \chi_f \rangle T_{i\to f} T^*_{i'\to f'} \tag{2}$$

and so on. To perform the complete experiment means to choose the detector states $\varepsilon, \varepsilon', \ldots$ and, possibly, to change the initial state ρ_0 in order to find all the independent amplitudes $T_{i\to f}$, relevant to the process under study, from the set of equations provided by the measurements of W, W', ... The total number of the amplitudes and therefore recipes for the complete experiment are model dependent.

2. Complete experiment in photoionization with excitation: General considerations

Consider now the photoionization

$$\gamma + A(\alpha_i J_i) \to A^+(\alpha J) + e^- , \tag{3}$$

where both the initial atomic and the final ionic states are characterized by sharp total electronic angular momenta, J_i and J, respectively. All other quantum numbers that are necessary for specifying the states are denoted by α_i and α. Within the classic treatment of the electromagnetic field with polarization vector $\boldsymbol{e}$, the expression for the transition operator in the long-wave dipole approximation is reduced to $T = \boldsymbol{eD} = \sum_\lambda e_\lambda D_\lambda$, where $\boldsymbol{D}$ is the atomic dipole momentum operator and $\lambda = 0, \pm 1$ denotes the spherical components of vectors. The basis set $|\xi_i\rangle = |\alpha_i J_i M_i\rangle$ is natural for the initial atomic state, where M_i is the projection of the angul ar momentum J_i. For the final state of process (3) the basis set $|\chi_f\rangle = |\alpha J, \ell j : J_t M_t\rangle$ is convenient in order to use directly the conservation of the total angular momentum and parity. Here ℓ and j are the orbital and total angular momenta of the photoelectron; J_t is the total angular momentum of the final state in (3), $\boldsymbol{J_t} = \boldsymbol{J} + \boldsymbol{j}$, and M_t is its projection. Due to the Wigner-Eckart theorem

$$\langle \alpha J, \ell j : J_t M_t | D_\lambda | \alpha_i J_i M_i \rangle = (2J_t + 1)^{-1/2} (J_i M_i, 1\lambda | J_t M_t) D(\ell j, J_t) \tag{4}$$

the dependence of the amplitudes on the projections M_i, M_t and λ is explicitly given by the Clebsch-Gordan coefficient $(J_i M_i, 1\lambda | J_t M_t)$. Measurement of the partial-wave amplitudes

$$D(\ell j, J_t) \equiv \langle \alpha J, \ell j : J_t || D || \alpha_i J_i \rangle \equiv d_{\ell j J_t} \exp i\delta_{\ell j J_t}, \tag{5}$$

which contain the dynamics of the photoionization, is the purpose of the complete photoionization experiment. In eq. (5) $d_{\ell j J_t}$ is the (non negative) absolute value of the amplitude and $\delta_{\ell j J_t}$ is its phase. Expressions of the type (1) and (2) for observable quantities in terms of the photoionization amplitudes (5) can be obtained by standard methods with the use of the statistical tensor formalism. Many examples are given in [2].

Discussions and implementations of the complete photoionization experiments started around 25 years ago [3]; more recent experimental and theoretical investigations of atomic photoionization have been motivated by the quest for such an experiment ([4-9] and references therein). Photoionization with only one active electron is normally considered a favorable object for a complete analysis due to a large probability of this process and relative simplicity of its theoretical description. In contrast, in photoionization with excitation (PIE), the absorption of a single photon leads to one electron being ionized and the second electron being excited. Since the electromagnetic interaction operator responsible for the photoionization is a single-particle operator, PIE is forbidden in the independent particle model and the occurrence of PIE is a consequence of electron-electron correlations. Studies by means of the complete experiment can provide very detailed information on such correlations and on the mechanisms of PIE.

The PIE shows up in photoelectron spectra as discrete satellites, described by different types of configuration interaction [10]. For example, the shake-up satellites can be considered as a result of final ionic state configuration interactions. Furthermore, at certain photon energies the satellites can be strongly enhanced by an autoionizing/Auger state decaying to the excited final ionic states; this resonance process is described in the framework of the final state configuration interactions between the discrete atomic state and the adjacent continuum [11]. The complete experiments require measurements of angle-differential and even spin-resolved cross sections which imply low counting rates. For this reason the complete experiment for the PIE has been realized to date only for the resonantly enhanced satellites. Since particular mechanisms of the photoionization are not implied in the amplitudes (5), the general approach to complete experiments for the resonant and direct PIE is the same.

The number of the independent amplitudes (5) for the process (3) is determined by the triangle rules, $\boldsymbol{J}+\boldsymbol{j}+\boldsymbol{J}_t=0$, $\boldsymbol{J}_i+\boldsymbol{J}_t+\boldsymbol{1}=0$, $\boldsymbol{\ell}+\boldsymbol{j}+\frac{1}{2}=0$, and conservation of parity, $(-1)^{\ell+1}\pi=\pi_i$, where π_i and π are the parities of the initial atomic and final ionic states, respectively. Complete experiments on the PIE so far have been performed with closed shell atoms ($J_i=0$, hence $J_t=1$ and J is a half integer). Below we consider this case. For atoms with $J_i=0$ the number of amplitudes (5) is two for final ionic states with $J=\frac{1}{2}$ and three for those with $J>\frac{1}{2}$. This leads to three and five independent real parameters, respectively, characterizing the absolute values of the amplitudes and their relative phases. One of the parameters can be excluded if we consider relative amplitudes instead of their absolute values; the latter are determined by the absolute integral cross section, which can be found separately. The independent dynamical parameters are extracted from observable quantities, as described in the next sections.

3. Angular distribution and spin polarization of photoelectrons

Angular distribution of photoelectrons gives a single independent dynamical parameter. For example, being produced by linearly polarized synchrotron radiation (SR), the angular distribution (at fixed photon energy) is given by

$$W(\theta)=\frac{W_0}{4\pi}\left[1+\beta P_2(\cos\theta)\right], \tag{6}$$

where W_0 is the integral cross section, θ is the angle between the directions of photoelectron emission and the electric field vector of the SR. The anisotropy parameter is expressed in terms of the amplitudes (5) as ([2], eq. (2.136))

$$\beta=(-1)^{J-1/2}\sqrt{6}\sum_{\ell\ell' jj'}\hat{\ell}\hat{\ell}'\hat{j}\hat{j}'(\ell 0,\ell' 0|20)\begin{Bmatrix}1 & 1 & 2\\ j & j' & J\end{Bmatrix}\begin{Bmatrix}\ell & \ell' & 2\\ j' & j & \frac{1}{2}\end{Bmatrix}\frac{\sqrt{\sigma_{\ell j}\sigma_{\ell' j'}}}{\sigma}\cos(\delta_{\ell j}-\delta_{\ell' j'})\,. \tag{7}$$

Here we used standard notations for the nj symbols and abbreviated $\hat{a} \equiv \sqrt{2a+1}$, put $D(\ell j, J_t = 1) = d_{\ell j} \exp i\delta_{\ell j}$, denoted the integral cross section of the process (3) by $\sigma = \sum_{\ell j} \sigma_{\ell j}$ with $\sigma_{\ell j}$ being the partial photoionization cross section into the ℓj continuum: $\sigma_{\ell j} = d_{\ell j}^2$.

The spin polarization component P_y of the photoelectrons normal to the reaction plane spanned by the linear polarization of the SR and the momentum of the photoelectron (the so-called 'dynamic' spin polarization) is of the form

$$P_y = \frac{\xi \sin 2\theta}{1 + \beta P_2(\cos\theta)}, \tag{8}$$

where ([2] eq.(2.155))

$$\xi = 3\sqrt{\frac{15}{2}} \sum_{\ell j \leq \ell' j'} (-1)^r \hat{\ell}\hat{\ell}'\hat{j}\hat{j}'(\ell 0, \ell' 0|20) \begin{Bmatrix} 1 & 1 & 2 \\ j & j' & J \end{Bmatrix} \begin{Bmatrix} \ell & \frac{1}{2} & j \\ \ell' & \frac{1}{2} & j' \\ 2 & 1 & 2 \end{Bmatrix} \frac{\sqrt{\sigma_{\ell j}\sigma_{\ell' j'}}}{\sigma} \sin(\delta_{\ell j} - \delta_{\ell' j'}), \tag{9}$$

with $r = \ell + j' + 1 - J$. The in-plane spin components of the photoelectron, which are described by two more parameters, A and α, are nonzero only for excitation with a circularly polarized photon beam. Only three of the four parameters β, ξ, A, α, are independent [8].

The values of β and P_y, providing two parameters, are sufficient to extract one absolute amplitude ratio and the relative phase of the two amplitudes in case of a final ionic state with $J = \frac{1}{2}$. In practice, however, spin polarization measurements suffer from poor energy resolution and low count rate, which result in unresolved photoelectron lines and large error bars [12]. For the case of $J > \frac{1}{2}$, a complete set of data, which allows for extraction of the photoionization amplitudes, cannot be obtained solely from the prop erties of the photoelectrons, even if all spin components of the photoelectrons are measured: a combination of different experimental techniques or a coincidence technique is required [8, 13].

4. Fluorescence polarimetry as a method to complete the photoionization experiment

A natural way to obtain the additional independent parameters for the PIE process is to study the polarization of the excited final ion state by detecting the angular distribution or the polarization of the ion fluorescence for discrete ion states or the ejected electrons for autoionizing/Auger ion states (see [6, 9, 14] and references therein), including coincidence photoelectron-fluorescence [5, 15] and photoelectron-Auger electron [7, 16] studies. The appropriate parameters are the alignment, $A_{20}(J)$, and orientation, $A_{10}(J)$, of the electronic angular momentum of the residual excited photoion. We concentrate on the measurements of these parameters by means of non-coincident fluorescence polarimetry. Fig. 1 shows one of the possible experimental schemes [9, 17]. The alignment parameter $A_{20}(J)$ determines the linear polarization, P_L, of fluorescence from the final ionic state. In the coordinate system of Fig. 1a, $A_{20}(J)$ can be extracted, for pure linearly polarized SR, from

$$P_L = \frac{W_{\parallel} - W_{\perp}}{W_{\parallel} + W_{\perp}} = \frac{3\kappa_2 A_{20}}{\kappa_2 A_{20} - 2}. \tag{10}$$

The circular polarization, P_C, of the fluorescence induced by circularly polarized SR (Fig. 1b) is proportional to the orientation parameter $A_{10}(J)$ and takes the form

$$P_C = \frac{W_+ - W_-}{W_+ + W_-} = \frac{\sqrt{3}\kappa_1 A_{10}}{\kappa_2 A_{20} + 1}. \tag{11}$$

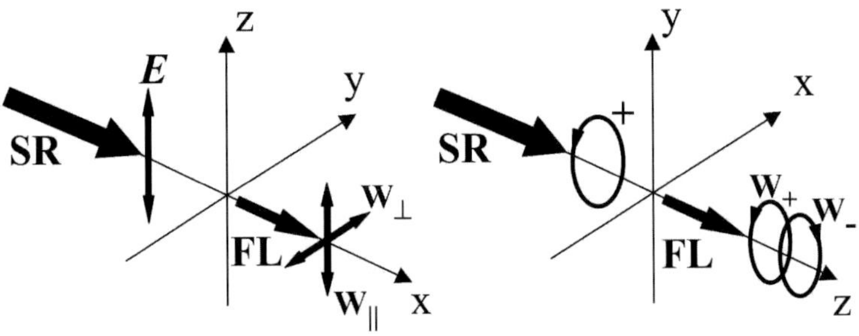

Figure 1. Geometry of the fluorescence polarimetry experiment.

The coefficient κ_k in eqs. (10) and (11),

$$\kappa_k = \sqrt{\frac{3}{2}}\hat{J}(-1)^{J+J'+k+1}\begin{Bmatrix} 1 & 1 & k \\ J & J & J' \end{Bmatrix} D_k(J), \tag{12}$$

depends on the angular momenta of the ion states before (J) and after (J') the radiative transition, while the factors $D_k(J)$ account for the depolarization of the electronic angular momentum J before the radiation takes place. A cross check of the fluorescence polarization data is possible: the parameters $A_{k0}(J)$ extracted from different fluorescence lines, originating from the same final photoion states, must be equal.

The parameters $A_{k0}(J)$ are expressed in terms of the photoionization amplitudes as

$$A_{k0}(J) = \sqrt{3}\hat{J}\,\rho^{\gamma}_{k0}\sum_{\ell j}(-1)^{J+j+k+1}\begin{Bmatrix} 1 & 1 & k \\ J & J & j \end{Bmatrix}\frac{\sigma_{\ell j}}{\sigma}, \tag{13}$$

where ρ^{γ}_{k0} are the statistical tensors of the SR photon (see table 1.1 of [2]). Unlike the photoelectron parameters (7) and (9), the photoion orientation and alignment are not related to the phase differences $\delta_{\ell j} - \delta_{\ell' j'}$ and only depend on the relative partial ionization cross sections $\sigma_{\ell j}/\sigma$.

High spectral resolution, which in general allows to resolve the fine-structure final ion states after the photoionization, the possibility of cross checking, and the possibility to deduce meaningful physical quantities (partial ionization propabilities) as a result of the measurements represent important advantages of the fluorescence polarimetry method. The major drawback of the method is the poorly controlled variation of the alignment and orientation of the photoion state due to additional population and depolarization of this state, taken into account by the factor $D_k(J)$ in eq. (12). For example, it is difficult, if possible at all, to measure quantitatively the effects of the possible radiative cascades. Calculation of the cascade is also problematic, since one generally needs to know the initial relative population of the ion levels after the photoionization, their polarization, and the probabilities of radiative transitions involed in the cascades. The inherent difficulties in separating the cascade effects stimulated correlation measurements between the photoelectron and subsequent polarized fluorescence photon as a method of the complete PIE experiment [5, 15]. Nevertheless, as shown below, the non-coincident fluorescence polarization method can also be used for this purpose.

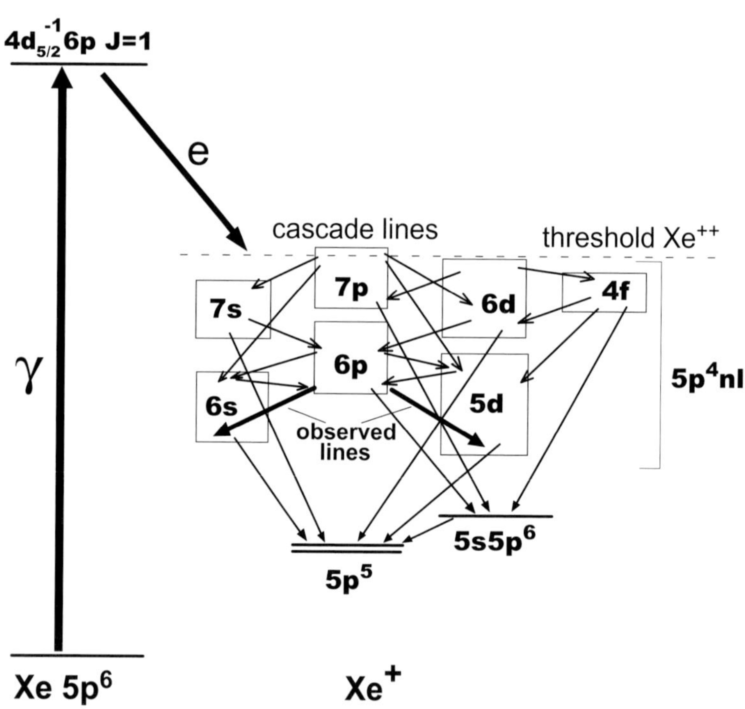

Figure 2. Scheme of transitions for PIE of Xe including radiative cascades in Xe^+.

5. Practical implementation of the fluorescence polarimetry method

As an illustrative example we consider ionization from the Xe $5p^6$ ground state into the excited Xe II $5p^4 6p$ ion states

$$\gamma_{SR} + \mathrm{Xe}\, 5p^{6\,1}S_0 \to \mathrm{Xe}^{+*}\, 5p^4(L_cS_cJ_c)6p[K]_J + e^-(\ell j) \tag{14}$$

in the region of the Xe* $4d^{-1}_{5/2}6p\,(J_r = 1)$ resonance at $h\nu$(SR) = 65.1 eV. The photoionization amplitudes $d_{\ell j}\exp i\delta_{\ell j}$ are different for each of the 21 fine-structure final ionic states $5p^4(L_cS_cJ_c)6p[K]_J$. Depending on J the contributing photoelectron partial waves are $\varepsilon s_{1/2}$, $\varepsilon d_{3/2}$ (for $J = \frac{1}{2}$); $\varepsilon s_{1/2}$, $\varepsilon d_{3/2}$, $\varepsilon d_{5/2}$ ($J = \frac{3}{2}$); $\varepsilon d_{3/2}$, $\varepsilon d_{5/2}$, $\varepsilon g_{7/2}$ ($J = \frac{5}{2}$); $\varepsilon d_{5/2}$, $\varepsilon g_{7/2}$, $\varepsilon g_{9/2}$ ($J = \frac{7}{2}$). Values of the angular anisotropy parameter β [eq. (7)] for all states of the Xe II $5p^4 6p$ configuration were determined in [18, 19]. The dynamical spin polarization P_y [eq. (8)] was measured in [12] for some of the final ionic states, but only for one of them (with $J = \frac{1}{2}$) it was possible to obtain the parameter ξ and to extract the relative magnitude of the PIE amplitudes of the $\varepsilon s_{1/2}$ and $\varepsilon d_{3/2}$ channels with reasonable accuracy.

In order to complete the PIE experiment, fluorescence polarimetry studies (see Fig. 1) were carried out at the ELETTRA storage ring in Trieste (Italy). At the "Circular Polarization" beamline XUV photons with a high degree of linear (>0.99) and circular (>0.98) polarization

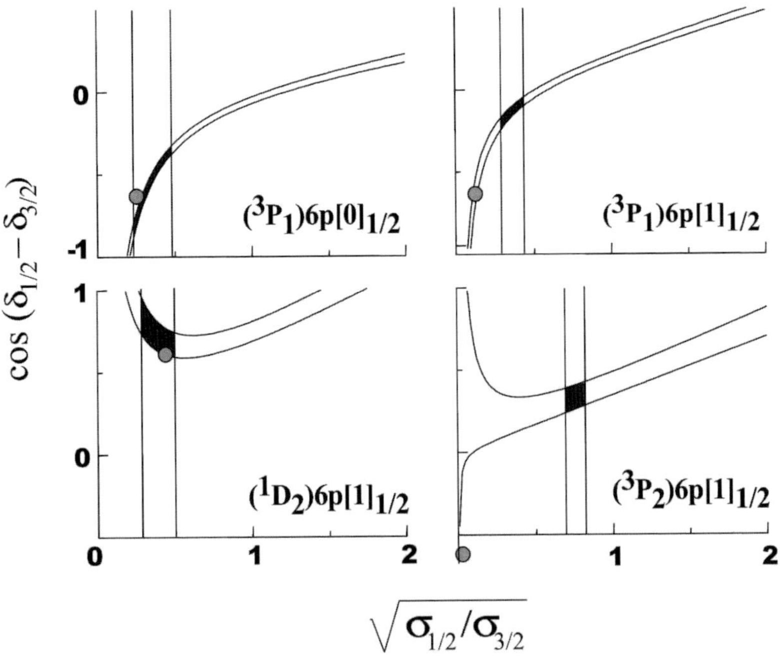

Figure 3. Results of a complete experiment for the PIE of Xe (14) into the Xe^+ $5p^46p$ states with J=1/2. Allowed values of the absolute ratio of the partial photoionization cross sections into the $\varepsilon s_{1/2}$ and $\varepsilon d_{3/2}$ channels and their phase differ ence, determined from experiment, are represented by the black areas. The grey circles show the theoretical results of [17].

can be used to excite the Xe* $4d^{-1}_{5/2}6p\,(J_r = 1)$ resonance. The fluorescence photons emitted from the interaction region between the effusive gas jet and the monochromatized SR were collected by a convex lens producing a parallel beam. A cone with opening angle of about 5° around the propagation direction of the SR is collected in this way. Before entering the high resolution spectrometer (Jobin Yvon HR460) for spectral analysis, the polarization of the fluorescence was analyzed by a sheet polarizer (linear polarization) and an additional quarter-wave plate (circular polarization) [9, 17]. Typical Xe gas pressures were about 5x10^{-5} mbar and typical acquisition times were about 30 minutes for one setting of the polarizer optics. More details of the experimental procedure are given in refs. [17, 20].

To extract the parameters A_{10}, A_{20} and to further obtain the relative partial cross sections from eq. (13), one needs to find the depolarization factors D_k in eq. (12). With respect to the experimental conditions, there are two depolarization effects that have to be taken into account: the hyperfine interactions, leading to the precession of the electronic angular momentum J around the total ionic angular momentum $\boldsymbol{F} = \boldsymbol{J} + \boldsymbol{I}$, where $\boldsymbol{I}$ is the nuclear spin, and the fluorescence cascades. Especially the treatment of the developed radiative cascades (Fig. 2) is very difficult. The key point allowing to simplify the task is the assumption of an "isotropic" radiative cascade population of the final Xe II $5p^56p$ ionic levels. The isotropic model, confirmed in our case experimentally, implies that the last line of the radiative cascades is unpolarized and therefore all magnetic substates of the final state in a cascade are equally populated by the cascade. In the isotropic model the depolarization factor can be expressed in a product form [20]: $D_k(J) = D(J)\,G_k(J)$. Here $G_k(J)$ is the depolarization factor due to the hyperfine interaction, calculated by standard methods [21] under the assumption that the PIE populates coherently the hyperfine levels of the final ionic state. The factor $D(J)$ is the relative probability to populate the final ionic state $A^+(\alpha J)$ by the PIE process, provided the radiative cascades also populate this state. A way to find $D(J)$ is to use the relative line intensities from the experimental photoelectron spectrum as input in the theoretical calculation of the fluorescence cascades. Necessary conditions for using this method are the availability of the photoelectron spectrum in a broad energy range with sufficient energy resolution, a reliable identification of the photoelectron lines, an accurate theoretical model to calculate the relative probabilities of the radiative transitions in the ion, and the possibility to account for all relevant pathways of the cascades. Such an analysis has been realized for the reaction (14) in [9, 20] by utilizing photoelectron spectroscopic data from [22] and extensive multiconfigurational intermediate-coupling calculations of the optical transition probabilities and the radiative cascades.

The combined data on A_{10} and β are sufficient to determine the absolute ratio $\sigma_{1/2} : \sigma_{3/2}$ and $\cos(\delta_{1/2} - \delta_{3/2})$ (we omit the index ℓ for brevity), as demonstrated in Fig. 3 for PIE into a few ion states with $J = \frac{1}{2}$. Combined data on A_{10}, A_{20}, β and ξ provide four independent quantities to find two absolute ratios and two phase differences of the three amplitudes, i.e. to resolve the problem of a complete experiment for the PIE into the ionic states with $J > \frac{1}{2}$. The statement that the four equations are independent is based on the fact that the corresponding Jacobian does not vanish [13, 17]. Only three parameters (A_{10}, A_{20}, and β) of the four quantities are currently available from experiments. Therefore, the particular solution for the amplitudes cannot be extracted, although the allowed amplitude space can be strongly reduced. Nevertheless, the values of the relative amplitudes can be fixed with the data available under further realistic assumptions (see Section 6).

6. Theoretical interpretation of the PIE complete experiment and discussion

For the strong resonance case, which is an excellent approximation for the above example in Xe, the theoretical description of the PIE can be further specified and reduced to the resonant Auger/autoionization decay to the excited ionic state. Consider ionization in the region of an isolated autoionizing state $|\,\alpha_r J_r\rangle$, characterized by the total angular momentum J_r and other quantum numbers α_r. Configuration interaction of the autoionizing state with the adjacent continuum is described by the matrix elements of the Coulomb interaction V,

$$V(\ell j, J_r) \equiv \langle \alpha J, \ell j : J_t \,||\, V \,||\, \alpha_r J_r \rangle\,. \tag{15}$$

Using Fano theory [11], as applied to the angular distribution and spin polarization of photoelectrons in the region of the isolated resonance, and considering the case of vanishing probability of the direct PIE, the photoionization amplitude takes the product form [23]

$$D(\ell j, J_r) = \langle \alpha_r J_r \,||\, D \,||\, \alpha_i J_i \rangle \frac{1}{E - E_r + i\Gamma/2} V(\ell j, J_r)\,. \tag{16}$$

Here $\Gamma = 2\pi \sum_{\ell j} |V(\ell j, J_r)|^2$ is the decay width of the resonance into the ionic state $|\,\alpha J\rangle$. Equation (16) symbolizes the two-step regime: the amplitude of the photoionization is factorized into a part describing the photoexcitation to the resonant state and a part describing the decay of this state. The Auger/autoionization decay amplitudes $V(\ell j, J_r)$ are now becoming the principal object of the complete experiment, because the first two factors are the same constants for all the partial amplitudes $D(\ell j, J_r)$. The complete experiment for the Auger decay in the two-step formulation is discussed in detail in ref. [24].

Fig. 3 compares the results of the complete PIE experiment to the ionic states with $J = \frac{1}{2}$ with the predictions obtained within the multiconfigurational Dirac-Fock approximation. The details of the theoretical model are described elsewhere [17]. Further improvements of the theory are needed to achieve better agreement with data from the complete experiment. The most important improvement of the model is expected to arise from a more precise description of the photoelectron continuum beyond the one-electron approximation.

For the states with $J > \frac{1}{2}$ a further step can be performed assuming that the relativistic splitting of the photoelectron continuum wavefunctions is negligible. This assumption is confirmed by the Dirac-Fock calculations, which show that the relative phase of continuum electron wavefunctions with the same ℓ, but different j, is less than 5° in all cases. Setting this difference to zero, the remaining three parameters can be found from experiment, i.e., the two absolute ratios and one phase difference. Some selected results are shown in Table 1. For the ionic states with $J = \frac{3}{2}$ the calculations give satisfactory results. Although the relative phase of the amplitudes is almost within the error bars, the data for the $J = \frac{7}{2}$ state on the line intensity and the branching ratios indicate a strong overestimate of the contribution from the $g_{9/2}$ channel, which points to an effect not included in the theoretical calculations. This demonstrates that only the complete data set can assess the validity of a theoretical model.

Table 1. Dynamic parameters for the PIE process (14) for a few selected final ionic states. The branching ratios $\sigma_j : \sigma$ are given in %; the relative photoline strengths $I_{\rm rel}$ are normalized to 100 when summed over the $5p^4 6p$ manifold.

Final ionic state		$\sigma_{1/2}:\sigma$	$\sigma_{3/2}:\sigma$	$\sigma_{5/2}:\sigma$	$\cos(\delta_s - \delta_d)$	$I_{\rm rel}$
$5p^4(^3P_2)6p[1]_{3/2}$	Exp	6(3)	16(3)	78(5)	-0.55(13)	15.2[a]; 15.05[b]
	Th	0.3	19.3	80.4	-0.65	12.45
$5p^4(^1D_2)6p[2]_{3/2}$	Exp	7(5)	49(7)	44(7)	0.07(38)	6.79[a]; 5.87[b]
	Th	1.7	46.2	52.1	-0.66	6.07
		$\sigma_{5/2}:\sigma$	$\sigma_{7/2}:\sigma$	$\sigma_{9/2}:\sigma$	$\cos(\delta_d - \delta_g)$	$I_{\rm rel}$
$5p^4(^1D_2)6p[3]_{7/2}$	Exp	12(9)	13(10)	75(10)	0.50(33)	2.44[a]; 3.66[b]
	Th	2.3	1.7	96.0	0.88	17.07

[a] Ref. [22], [b] Ref. [19]

Acknowledgments

We wish to thank P. O'Keeffe and S. Fritzsche for collaboration. ANG gratefully acknowledges financial support of the RFBR under grant 04-02-17236.

References

[1] Bederson B 1969 *Comm. At. Mol. Phys.* **1** 41
Kleinpoppen H 1971 *Phys. Rev.* A **3** 2015
Eminyan M, MacAdam K, Slevin J and Kleinpoppen H 1973 *Phys. Rev. Lett.* **31** 576

[2] Balashov V V, Grum-Grzhimailo A N and Kabachnik N M 2000 *Polarization and Correlation Phenomena in Atomic Collisions: A Practical Theory Course* (New York: Academic / Plenum Publishers)

[3] Duncanson Jr J A, Strand M P, Lindgard A and Berry R S 1976 *Phys. Rev. Lett.* **37** 987
Hansen J C, Duncanson Jr J A, Chien R L and Berry R S 1980 *Phys. Rev.* A **21** 222
Cherepkov N 1979 *J. Phys. B: At. Mol. Phys.* **12** 1279
Heinzmann U 1980 *J. Phys. B: At. Mol. Phys.* **13** 4367
Heckenkamp Ch, Schäfers F, Schönhense G and Heinzmann U 1984 *Phys. Rev. Lett.* **52** 421
Kaminski H, Kessler J and Kollath K J 1980 *Phys. Rev. Lett.* **45** 1161
Kessler J 1981 *Comm. At. Mol. Phys.* **10** 47
Klar H and Kleinpoppen H 1982 *J. Phys. B: At. Mol. Phys.* **15** 933

[4] Becker U 1998 *J. Electr. Spectrosc. Relat. Phenom.* **96** 105
Becker U and Langer B 1998 *Phys. Script.* **T78** 13
Godehusen K, Zimmermenn P, Verweyen A, von dem Borne A, Wernet Ph and Sonntag B 1998 *Phys. Rev.* A **58** R3371
Prümper G, Zimmermann B, Becker U and Kleinpoppen H 2001 in *Complete Scattering Experiments* edited by U Becker and A Crowe (New York: Kluwer Academic / Plenum Publishers), p 141
Snell G, Hergenhahn U, Müller N, Drescher M, Viefhaus J, Becker U and Heinzmann U 2001 *Phys. Rev.* A **63** 032712

[5] West J B, Ross K J and Beyer H J 1998 *J. Phys. B: At. Mol. Opt. Phys.* **31** L647
Ueda K, West J, Ross K J, Beyer H-J and Kabachnik N M 1998 *J. Phys. B: At. Mol. Opt. Phys.* **31** 4801

[6] Lörch H, Bizau J M, Scherer N, Diehl S, Cubanes D, Zerouni O, Wuilleumier F J, Schmidt V and Johnson W R 1999 *J. Phys. B: At. Mol. Opt. Phys.* **32** 2215

[7] DeFanis A, West J B, Ross K J, Ueda K, Beyer H-J, Amusia M Ya and Chernysheva L V 1999 *J. Phys. B: At. Mol. Opt. Phys.* **32** 5739

[8] Schmidtke B, Drescher M, Cherepkov N A and Heinzmann U 2000 *J. Phys. B: At. Mol. Opt. Phys.* **33** 2451

[9] O'Keeffe P, Aloïse S, Meyer M and Grum-Grzhimailo A N 2003 *Phys. Rev. Lett.* **90** 023002

[10] Schmidt V, 1997 *Electron Spectrometry of Atoms using Synchrotron Radiation* (Cambridge: University Press)

[11] Fano U 1961 *Phys. Rev.* **124** 1866
Fano U and Cooper J W 1965 *Phys. Rev.* **137** A1364

[12] Hergenhahn U, Snell G, Drescher M, Schmidtke B, Müller N, Heinzmann U, Wiedenhöft W and Becker U 1999 *Phys. Rev. Lett.* **82** 5020

[13] Cherepkov N A and Semenov S K 2004 *J. Phys. B: At. Mol. Opt. Phys.* **37** 1267

[14] Hausmann A, Kämmerling B, Kossmann H and Schmidt V 1988 *Phys. Rev. Lett.* **61** 2669
Hamdy H, Beyer H-J, West J and Kleinpoppen H 1991 *J. Phys. B: At. Mol. Opt. Phys.* **24** 4957
Schill R H et al 2003 *J. Phys. B: At. Mol. Opt. Phys.* **36** L57

Wetzstein W, Bencheckmoumou T, Zatsarinny O I, Grum-Grzhimailo A N, Bartschat K and Mehlhorn W 2003 *J. Phys. B: At. Mol. Opt. Phys.* **36** 3961

[15] Beyer H-J, West J B, Ross K J, Ueda K, Kabachnik N M, Hamdy H and Kleinpoppen H 1995 *J. Phys. B: At. Mol. Opt. Phys.* **28** L47

[16] Kämmerling B and Schmidt V 1993 *J. Phys. B: At. Mol. Opt. Phys.* **26** 1141
Scherer N, Scharphorst S J and Schmidt V 2001 in *Complete Scattering Experiments* edited by U Becker and A Crowe (New York: Kluwer Academic / Plenum Publishers), p 155
Ueda K, Shimizu Y, Chiba H, Sato Y, Kitajima M, Tanaka H and Kabachnik N M 1999 *Phys. Rev. Lett.* **83** 5463

[17] O'Keeffe P, Aloïse S, Fritzsche S, Lohmann B, Kleiman U, Meyer M and Grum-Grzhimailo A N 2004 *Phys. Rev.* A **70** 012705

[18] Aksela H, Jauhiainen J, Nõmmiste E, Sairanen O-P, Karvonen J, Kukk E and Aksela S 1996 *Phys. Rev.* A **54** 2874

[19] Langer B, Berrah N, Farhat A, Hemmers O and Bozek J D 1996 *Phys. Rev.* A **53** R1946

[20] Meyer M, Marquette A, Grum-Grzhimailo A N, Kleiman U and Lohmann B 2001 *Phys. Rev.* A **64** 022703

[21] Fano U and Macek J H 1973 *Rev. Mod. Phys.* **45** 553
Greene C H and Zare R N 1982 *Phys. Rev.* A **25** 2031

[22] Aksela H et al 1995 *Phys. Rev.* A **51** 1291

[23] Kabachnik N M and Sazhina I P 1976 *J. Phys. B: At. Mol. Phys.* **9** 1681
Kabachnik N M, Sazhina I P and Ueda K 1999 *J. Phys. B: At. Mol. Phys.* **32** 1769

[24] Kabachnik N M 2004 *J. Electr. Spectrosc. Relat. Phenom.* **137-140** 305

Inst. Phys. Conf. Ser. No. 183
Paper presented at Int. Conf. Electron and Photon Impact Ionization and Related Topics, Louvain-la-Neuve, 2004

Double photoionisation of molecules via TOF electron-electron coincidence experiments

J.H.D. Eland
Physical and Theoretical Chemistry Laboratory, Oxford, UK

Abstract. This paper reports the results of TOF-PEPECO experiments on the polyatomic molecules ethylene, butadiene, *cyclo*octatetraene, benzene and naphthalene. New values are obtained for the double ionisation energies and some light is cast on the direct and indirect double ionisation mechanisms in photoionisation.

1. Introduction

The time-of-flight photoelectron-photoelectron coincidence technique (TOF-PEPECO) [1] provides a new approach to the spectra of doubly charged cations, and also reveals the electron distributions produced in their formation by single photon ionisation. The spectra of several atoms and small molecules have already been reported [1-6]; this paper gives new results on the spectra obtained for some larger molecules where vibrational structure is generally not resolved. The results are compared, where possible, with previous measurements by the TPEsCO (threshold photoelectrons in coincidence) technique and other methods.

The special advantages of the TOF-PEPECO technique are that photoelectrons are collected from almost the whole 4π solid angle and that electrons of all relevant energies are collected without energy discrimination. The lack of discrimination by energy or angle of emission means that the spectra are *complete*, in the sense that they properly represent the population of doubly charged states created by photoionisation. The experiments are less demanding than many coincidence techniques because of the very high collection efficiency. This paper also points out for the first time that in addition to the spectra and electron distributions, the technique gives the quantum yield of double ionisation from notional or real intermediate singly ionised states. For single ionisation, the quantum yield of ionisation efficiency is defined as the number of ions produced, on average, for each photon of sufficient energy absorbed. The quantum yield of double ionisation is defined analogously as the number of double ionisations produced following each single ionisation which leaves the (mono)cation with enough energy to ionise again.

2. Experimental method

The technique has now been described several times [2-6], so only a brief outline is given here. Pulses of ionising light are created in a capillary gas discharge lamp, by switching the charge stored in a capacitor (150 pF, 10 kV) to the lamp cathode using a fast hydrogen thyratron. Individual wavelengths are selected in a toroidal grating grazing incidence monochromator and refocused to intersect an effusive jet of target gas. Ionisation happens in a divergent magnetic field provided by a shaped permanent magnet, which closes one end of the magnetic bottle formed by a long (5.5 m) solenoid. The electrons guided along the solenoid field lines are detected at a chevron microchannel plate detector after traversing the long flight path. A small electric potential difference applied between the source region and the flight tube ensures that electrons formed with zero initial energy reach the detector within about 20 μs of the light pulse. Electron arrival times are recorded by a multi-hit time-to-digital converter and can be converted to electron energies. The light pulse length is between 5 ns and 10 ns, and determines the energy resolution for electron energies above about 10 eV. For lower energies the resolution is determined mainly by the magnetic field strengths, which fix the degree to which the electron trajectories are made parallel [7]. For most dication spectra at a photon energy of 40.814 eV, (HeI), and wherever the highest electron energy is less than 10 eV, the resolution is 100 meV or less. For electron energies below 3 eV, peak widths are 30 meV or better. Energy calibration is by comparison with the known single ionisation spectra of oxygen or xenon recorded before and after each experimental run. Absolute calibration of the overall electron collection and detection efficiency, *f*, is by comparison of measured single and pair electron signals with the known single and double photoionisation cross-sections for rare gas atoms. The efficiency is very high, 50 to 60% depending on the condition of the detector.

Once the efficiency, *f*, is known, the quantum yield of double ionisation is easily derived from the numbers of electron pairs *P* and single electrons *S* detected, with a common energy for the single electron and the higher energy electron of the pair.

$$P = N_{2+}f^2$$

$$S = N_+f + N_{2+}(1-f)f$$

Here N_+ and N_{2+} are the true rates of single and double ionisation and their ratio is the quantum yield. In double ionisation to a single final state, the energies of the two electrons released have a mirror image relationship about half the maximum available energy. If one electron has the maximum energy the other has zero, or both may have half the maximum, for instance. Because of this, quantum yields can be determined correctly using the above simple equations only up to halfway from the double ionisation threshold to the photon energy in use.

3. Results

The molecules are discussed separately in the following subsections, but the first three spectra are presented together in Fig. 1. Ethylene (ethene), *trans* 1-3 butadiene and cyclooctatetraene illustrate the trend towards more and more featureless spectra as molecules get larger. Benzene and naphthalene are discussed and presented in more detail, as these aromatic molecules show characteristic features. Although complete two-electron distributions and quantum yields are measured every time, only a few are presented here as most are rather featureless, or show general features which are well characterised by typical examples.

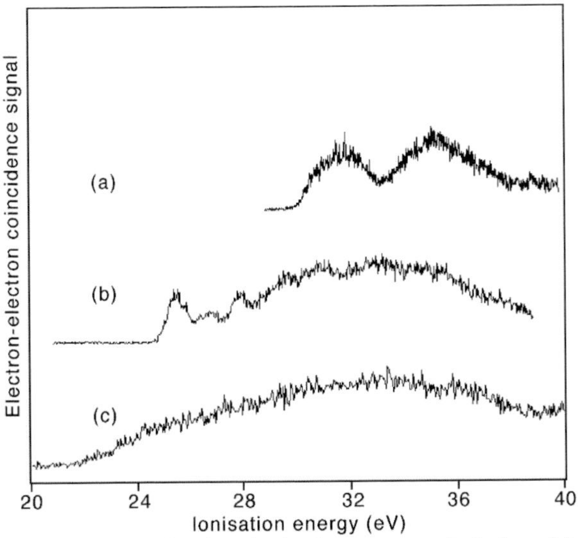

Figure 1. Double photoionisation spectra of ethylene (a), butadiene (b) and cyclooctatetraene (c)

3.1. Ethylene

Ethylene has only two more atoms than acetylene, whose TOF-PEPECO and TPEsCO double ionisation spectra are richly structured [8]. Nevertheless, as is seen in Fig 1(a), ethylene's dication spectrum is very bland. There are two main bumps, with a first main onset near 29.95 eV and a weak ramp, barely distinguishable from noise, starting near 29.5 eV. There is apparently a single resolved vibration line at 31.78 eV, which needs further confirmation, but the rest of the scatter is statistical noise. Detailed theoretical calculations on the states of the $C_2H_4^{2+}$ ion have been carried out by Ohrendorf et al. [9], partly in an effort to interpret the Auger spectrum. The paper actually called for PEPECO measurements of exactly the present kind, but had an optimistic view of the structure that would be revealed. These calculations and others [10] have shown that the lowest energy state of the ethylene dication is not planar, but has a 90° twisted D_{2d} geometry. The C-C bond length is also calculated to be significantly different from that in the neutral molecule. Thus ionisation to the predicted ^1A_g ground state will be accompanied by strong excitation of at least the torsional vibration and C-C stretch. The Ohrendorf et al. calculations predict even more complicated vibronic structure, because of coupling between three low lying electronic states. Twenty electronic states of the dication are predicted to lie between the threshold and 40 eV, without consideration of the alternative isomeric structures [10] which may be accessible. Thus there are ample grounds for explaining the lack of observed structure in the spectrum as a result of congestion.

3.2. Butadiene

The double ionisation spectrum of trans 1-3 butadiene in Fig 1(b) shows no trace of resolved vibrational structure. There are three bumps with maxima at 25.4, 26.7 and 27.8 eV; the first onset at 24.76 ±.1 eV is rather sharp. The peak positions are in good agreement with the results of a detailed study by Auger spectroscopy, double charge transfer (DCT) spectroscopy and theoretical methods [11]. According to the photoelectron spectrum and its interpretation [12] there is a 2.5 eV energy gap between the two outermost occupied π-orbitals, $1b_g$ and $1a_u$, so we might expect the ground state of the dication to be a singlet. This idea is also confirmed by the Auger and DCT studies, which show that the state at 25.4 eV is indeed a singlet. The three clear features in the present spectrum are better resolved than in the Auger and DCT spectra, but show no trace of vibrational structure. As many states are predicted, spectral congestion may be the main reason again. Another contributory factor is likely to be rapid dissociation, as PEPIPICO (photoelectron-photoion-photoion coincidence) experiments in this laboratory show strong formation of ion pairs, with no sign of a stable or metastable parent dication.

3.3. *Cyclooctatetraene*

The loss of two π-electrons from *cyclo*-C_8H_8 leaves a structure which could be aromatic and might become planar. Calculations [10] suggest however, that the planar structure may not be the lowest energy isomer of $C_8H_8^{2+}$. In any case there will be a large change in geometry between the neutral molecule (a puckered ring) and the dication. Together with the large number of electronic states to be expected, this could account for the complete lack of structure in the double ionisation spectrum, Fig 1(c). Again, this is not a particularly stable molecule, and rapid dissociation is also probable. Unfortunately the double photoionisation spectrum of this compound is rather typical of those of many large molecules.

3.4. *Benzene*

Because a doubly charged ion, $C_6H_6^{2+}$ is observed in the mass spectrum [13], we know that at least some benzene dications live long enough to support resolved vibrational states. Theoretical calculations [14-17] indicate that the three lowest states of the dication should be of $^3A_{2g}$, $^1E_{2g}$ and $^1A_{1g}$ symmetry, in that order, arising from removal of two electrons from the outermost e_{1g} orbital of the parent molecule. The experimental evidence from Auger spectra [18] and double charge transfer (DCT) spectra [19,20] placed the first triplet near 24.6 eV and a singlet state near 26 eV, and this analysis was largely confirmed by a TPEsCO study [21]. The TPEsCO spectrum showed weak vibrational structure, interpreted as a progression in the ring breathing mode, below a broad peak at 25.3 eV. These two features are confirmed by the present TOF-PEPECO spectra shown in Fig. 2. The first definite resolved peak is at 24.68 eV and a second one is found 120 meV above. There are weaker features at lower apparent energy including a possible peak at 24.44 eV, but these may be due to a contribution from the higher energy line (26.9 eV) in the 461X doublet from NeII, used to obtain the spectrum. Spectra have been taken at photon energies of 26.8, 40.8 and 48.4 eV. The spectra show additional broad peaks at 27.4 and 30 eV, followed by a rather smooth distribution of intensity right up to 48 eV. The peak at 27.4 eV probably corresponds to the triplet state(s) found in the DCT spectrum at 27.3 eV. Because the theoretical calculations predict many states in this energy region, a more exact identification is not yet possible.

The double ionisation energy of benzene has been reported from electron impact mass spectrometry as 26 eV, and from photon impact as 25.3 eV [22]. That these values are substantially higher than the lowest double ionisation energies found here suggests that these onset techniques respond mainly to dication states formed with high probability.

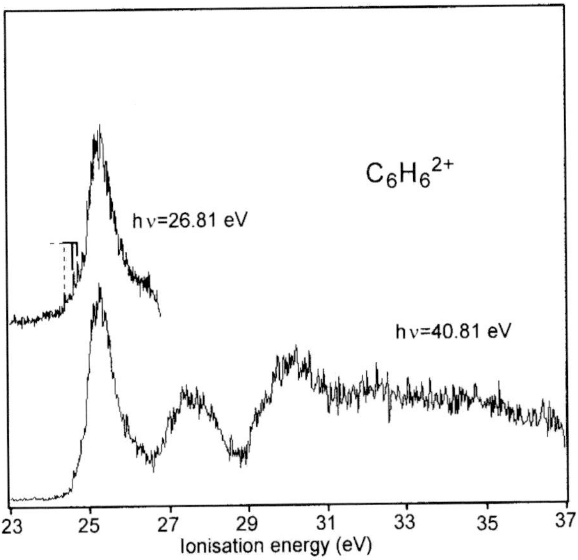

Figure 2. TOF-PEPECO spectra of benzene at photon energies of 40.81 (HeII) and 26.81 eV (NeII)

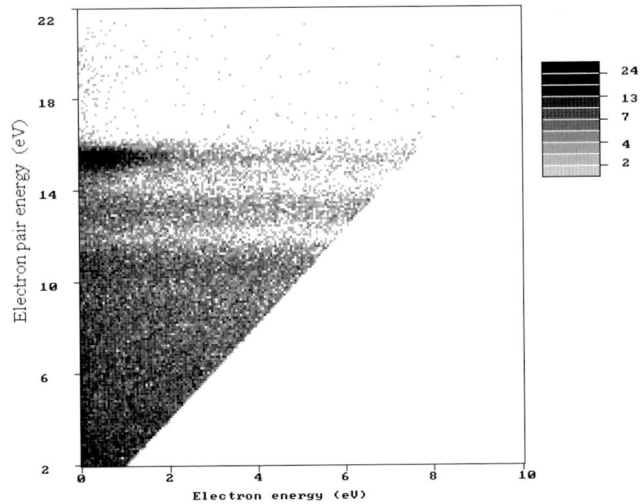

Figure 3. Electron distribution map for double photoionisation of benzene at 40.8 eV . In the complete distribution, each horizontal line is doubled in length by reflection at the point where it meets the diagonal.

An electron pair distribution map for benzene at 40.8 eV is given as Fig. 3. It shows that in formation of the lowest energy state(s) of the dication, near 25 eV, an uneven distribution of electron energy is favoured, with one electron near zero and the other near the maximum energy. This type of distribution for the lowest energy dication states is prevalent for larger molecules, and a possible explanation for it is encapsulated in Fig. 4. The single ionisation spectrum (central curve) contains a weak unassigned feature, presumably a CI band, about 1 eV above the double ionisation threshold. The quantum yield of double ionisation (upper curve) is much less than 100% and shows a peak coinciding with the CI band. The superexcited $C_6H_6^+$ ions formed in the CI state may partly dissociate, but partly autoionise to the accessible dication state(s). Because CI bands gain intensity by mixing with one-electron allowed transitions, they are generally more prominent than doubly excited Rydbergs, and autoionise more effectively because their valence character gives better orbital overlap. It is in large molecules, where the double ionisation threshold is low, that such CI bands are found at energies where autoionisation is possible.

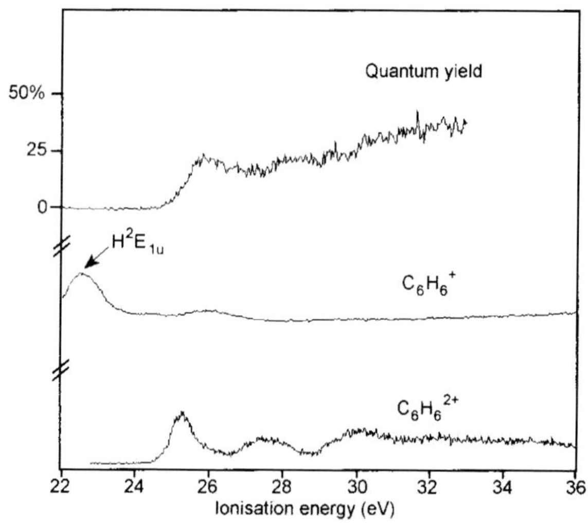

Figure 4. Double (lowest curve) and single (middle) photoionisation spectra of benzene and the yield of double ionisation at 40.81 eV.

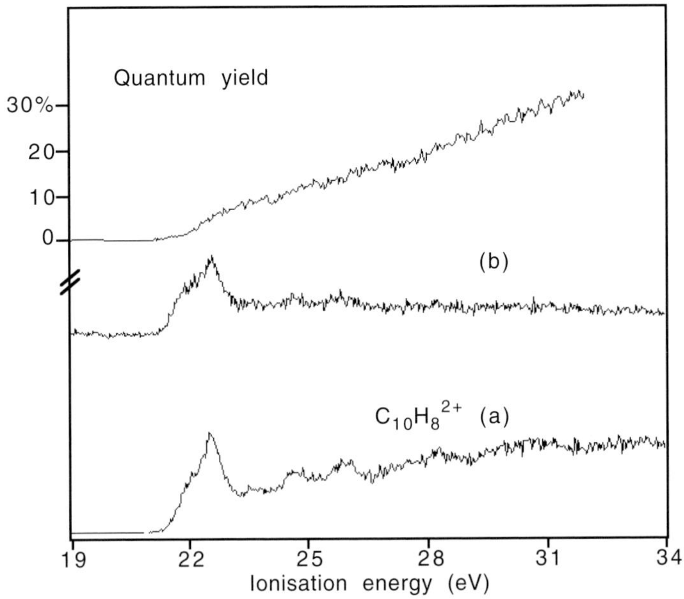

Figure 5. Complete double photoionisation spectrum of naphthalene at 40.8 eV (a) and the selected spectrum (b) where one electron has energy less than 1 eV. The upper curve shows the ionisation yield.

3.5. Naphthalene

Two forms of the double photoionisation spectrum of naphthalene are displayed in Fig. 5, together with the quantum yield. Because the electron pair distribution map for naphthalene has the same form as for benzene, selection of electron pairs including one low energy (<1 eV) electron emphasises the low-lying dication states. The quantum yield is lower in this case and shows no clear feature corresponding to a CI band; nevertheless there are CI states available, and the same mechanism probably applies. The onset of the spectrum is rather gradual, but a clear signal is present at 21.3±.1 eV, in good agreement with the onset determined by photoionisation mass spectrometry as 21.5±.5 eV [22]. There are peaks at 22, 22.5, 23.6, 24.6, 25.9 and 28.3 eV before a continuum sets in at higher energies. Because stable $C_{10}H_8^{2+}$ ions exist, there is a possibility that future higher resolution work will reveal vibrational structure in the lowest energy bands.

4. General conclusions

For most molecules larger than triatomic or tetraatomic, experience suggests that mainly unresolved double photoionisation spectra will be obtained. The polynuclear aromatic hydrocarbons may possibly be exceptions. For both small and large molecules the quantum yields of double ionisation as defined in this paper are substantially less than unity, showing that superexcited states of the monocations, exist, are formed by photon impact and mostly dissociate to singly ionised products rather than autoioinising.

References

[1] Eland J H D, Vieuxmaire O, Kinugawa T, Lablanquie P, Hall R I and Penent F 2003 *Phys. Rev. Lett.,* **90** 053003-7
[2] Eland J H D, Ho S S W and Worthington H L 2003 *Chem. Phys.* **290** 27-34
[3] Eland J H D 2003 *Chem. Phys.* **294** 171-186
[4] Hochlaf M and Eland J H D 2004 *J. Chem. Phys.* **120** in press
[5] Pilcher-Clayton A and Eland J H D 2004 *J. Electron Spectrosc.* in press
[6] Kinugawa T, Lablanquie P, Penent F, Paladoux J and Eland J H D 2004 *J Electron Spectrosc.* in press
[7] Kruit P and Read F H 1983 *J. Phys* **E16** 313-325
[8] Kinugawa T, Lablanquie P, Penent F, Paladoux J and Eland J H D 2004 *J. Electron Spectrosc.* in press
[9] Ohrendorf E, Koeppel H, Cederbaum L S, Tarantelli F and Sgamellotti A 1989 *J.Chem. Phys.* **91** 1734-53
[10] Lammertsma K, von Rague Schleyer P and Schwarz H 1989 *Angew. Chemie Int.Ed. Engl.* **28** 1321-1341
[11] Liegener C-M, Naves de Brito A, Agren H, Correia N, Griffiths W J, Svensson S and Harris F M 1992 *Phys. Rev.* **B46** 11295-11309
[12] Beez M, Bieri G, Bock H and Heilbronner E 1973 *Helv. Chim. Acta* **56** 1028-34
[13] Hustrulid A, Kusch P and Tate J T 1937 *Phys. Rev.* **52** 843-50
[14] Lammertsma K and Schleyer P 1983 *J. Amer. Chem. Soc.* **105** 1049-60
[15] Dewar M J S and Holloway M K 1984 *J. Amer. Chem. Soc.* **106** 6619-25
[16] Tarantelli F, Sgamellotti A, Cederbaum L S and Schirmer J 1987 *J. Chem Phys.* **86** 2201-15
[17] Krogh-Jespersen K 1991 *J. Amer. Chem. Soc.* **113** 417-25
[18] Spohr R, Beckmark T, Werme L O, Karlsson L, Nordling C and Siegbahn K 1970 *Physica Scripta* **2**, 31-53
[19] Appell J, Durup J, Fehsenfeld F C and Fournier P G 1974 *J. Phys.* **B7** 406-13
[20] Griffiths W J and Harris F M 1991 *Chem. Phys.* **157** 299-307
[21] Hall R I, Avaldi L, Dawber G, McConkey A G, MacDonald M A and King G C 1994 *Chem. Phys.* **187** 125-135
[22] Tobita S, Leach S, Jochims H W, Ruehl E, Illenberger E and Baumgaertel H 1994 *Can. J. Phys.* **72** 1060-1069
[23] Ruehl E, Price S D and Leach S 1989 *J. Phys. Chem.* **93** 6312-6321

Inst. Phys. Conf. Ser. No. 183
Paper presented at Int. Conf. Electron and Photon Impact Ionization and Related Topics, Louvain-la-Neuve, 2004

Simultaneous versus sequential two-photon double electron ejection from He^-

M. S. Brouri and X. Urbain

Laboratoire de physique atomique et moléculaire, Université catholique de Louvain, chemin du cyclotron 2, B-1348 Louvain-La-Neuve, Belgium

Abstract. The 1+1 photon resonance-enhanced double ionisation of He^- has been observed for the first time in the interaction of a fast ion beam with a pulsed laser beam between 375 and 425 nm. Although a clear enhancement of the He^+ yield is observed above the $1s3s\ ^3S$ threshold, a minimum is found at the exact location of all three photodetachment resonances located in this range. This pleads against the sequential ionisation model. The $1s3s4s\ ^4S$ resonance located below the $1s3s$ threshold appears to be ideally suited for the demonstration of electron correlation effects accompanying their simultaneous ejection. A careful analysis of the spectrum is needed, as spurious resonances existing in the 1+1 REMPI of $He(2^3S)$ and $He(2^3P)$ add up to the total ion yield.

1. Introduction

The negative helium ion is among the simplest few-electron systems, yet it possesses many interesting features that deserve special attention. With a fully correlated ground state of dominant configuration $1s2s2p\ ^4P$, He^- constitutes an ideal playground for the study of electron correlation and laser-induced collective phenomena. All excited states are doubly-excited and autodetaching, and pioneering studies have dealt with the measurement of partial and total photodetachment cross sections [1, 2, 3, 4]. Hanstorp has developed a high resolution apparatus for the precise study of photodetachment, based on laser-ion collinear geometry [5]. Partial cross sections have been measured for the three resonances located in the vicinity of the $He(1s3l\ ^3L)$ threshold by Klinkmüller *et al.* [3, 4].The partial cross sections are obtained by probing the final state with a second laser that excites the atom to a Rydberg state which is field-ionised, making the detection process highly selective. Similar experiments have recently revealed a rich structure of double Rydberg series in He^- [6]. Other recent experiments have investigated the triply-excited states with synchrotron radiation [7]. Extensive calculations have been performed by Xi and Froese Fischer [8], Ramsbottom and Bell [9], Brandefelt and Lindroth [10] and are still under way. To our knowledge however, there exists no experiment nor calculation dealing with He^- in strong fields. We have designed a unique experiment on He^-, where the 1+1 photon resonance-enhanced double ionisation can be observed for the first time.

2. Experiment

The weak binding energy (77.52 meV [11]) of the He^- ion forbids its production with regular ion sources. Its metastability ($\tau \sim 365$ μs for the $J = 5/2$ component [12]) however allows conventional beam transport and detection techniques. The apparatus is sketched in figure 1. A 4keV beam of He^+ ions is produced in a duoplasmatron source, that is converted into He($1s2s\ ^{1,3}S$) and $He^-(1s2s2p\ ^4P)$ in a Cs vapor cell. The production of He^- requires a large target density for multiple collisions to occur, resulting in 20% neutralization efficiency and 1% conversion into He^-. A double deflection clears the beam from its neutral and positive components before it enters a region pumped to high vacuum (10^{-8} mbar). However, some contamination of the ion beam by neutrals at the percent level still occurs due to the spontaneous detachment of the $J = 1/2$ and $J = 3/2$ components, that have a lifetime of the order of 11μs [12].

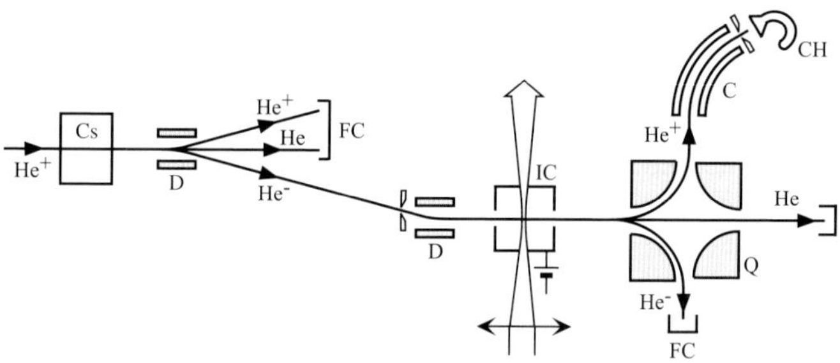

Figure 1. Cs: cesium vapour cell; D: parallel plate deflectors; FC: Faraday cups; IC: biased interaction cell; Q: quadrupolar deflector; C : cylindrical deflector; CH: channel electron multiplier.

The ion beam is illuminated at right angle by near-UV laser pulses produced by frequency-mixing the output of a dye laser operating with sulforhodamine, DCM and pyridine 1, with the Nd-YAG fundamental frequency to cover the range 375-425 nm. The laser pulse energy is measured by a pyroelectric detector coupled with a fast photodiode (*Ophir*) and the wavelength calibration relies on a double-etalon wavemeter (*Burleigh*). The laser beam is focused down to 20 μm, producing a peak intensity of 3×10^{11} W cm^{-2} for 10 mJ pulses. The ion beam cross section has a 1 mm diameter, which implies some spatial averaging of the laser intensity. Furthermore, 4 keV He^- ions travel through the laser spot (twice the waist) within 90 ps, a much shorter time than the pulse duration (5 ns).

The interaction takes place in a biased region, that accelerates the negative ions at the entrance and the positive ions at the exit. Subsequent energy analysis performed by a quadrupolar deflector [13] coupled with a 60° cylindrical deflector ensures the selection of laser-induced double ionisation events, while negative ions are collected at the opposite side of the quadrupole in a Faraday cup. Neutrals are collected in straight line by a Faraday cup, which may be replaced by a channel electron multiplier when operating with very low ion currents. In addition to the energy analysis, a straightforward time-of-flight selection of the events is performed with respect to the laser pulses, since all ions arrive at the detector (channel electron multiplier) within 10 ns.

As a drawback of this temporal clustering of events, we can not distinguish between single and multiple events occurring within the same laser shot. We have to apply a correction to the rate actually measured to account for the Poisson probability distribution :

$$N_{corr} = -N_p \, ln(1 - N_{meas}/N_p) \tag{1}$$

where N_{meas} (N_{corr}) is the measured (corrected) number of events recorded after N_p laser pulses. The corrected yield is normalized to the ion beam intensity to build the double ionisation spectra presented below. The measurements are essentially background free.

As already mentioned, the spontaneous detachment of the $J = 1/2$ and $J = 3/2$ ions produces a sizeable flux of neutrals, which prevents the detection of photodetached atoms. One way to circumvent this limitation is to operate with a pulsed ion beam of reduced intensity. The ion bunches are made by rapidly sweeping the beam in front of the 1 mm collimator with the help of a parallel plate condenser fed with a pulsed voltage. Typical bunch lengths of 10-15 ns are produced, that are brought in coincidence with the laser pulse. He atoms are detected in straight line with a channel electron multiplier. The bias voltage applied to the interaction region is raised to 500 volts to accelerate the ions entering the cell, thereby producing a time separation of the bunch of neutrals created there. A signal-to-noise ratio better than 10:1 has been achieved in this way.

3. Results and discussion

Spectra have been recorded between 375 and 425 nm at different laser pulse energies, ranging from 1 to 10 mJ. A typical spectrum is shown in figure 2. Sharp resonances appear superimposed on a structured background, which rises abruptly above 2.977 eV. This energy corresponds to the He($1s3s\ ^3S$)+e$^-$ detachment threshold. That state opens up the sequential ionisation process, the first photon detaching one electron from He$^-$ and leaving the atom in a $1s3l\ ^3L$ state, the second photon ionising it, as depicted by figure 3.

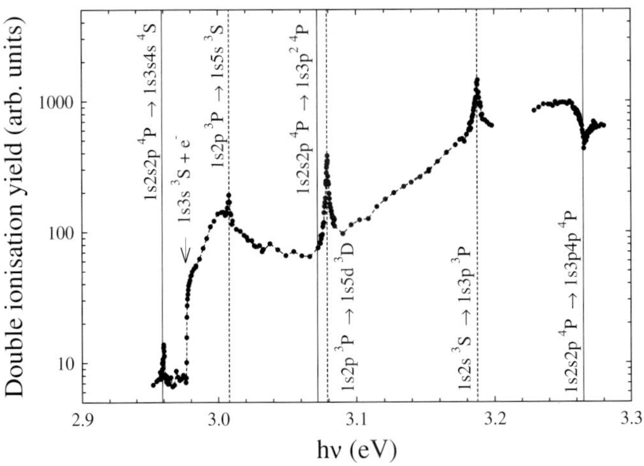

Figure 2. Double ionisation signal measured with 8.5 mJ laser pulses. Data have been corrected for saturation. Full lines: He$^-$ transitions ; dashed lines: He transitions; arrow: detachment threshold.

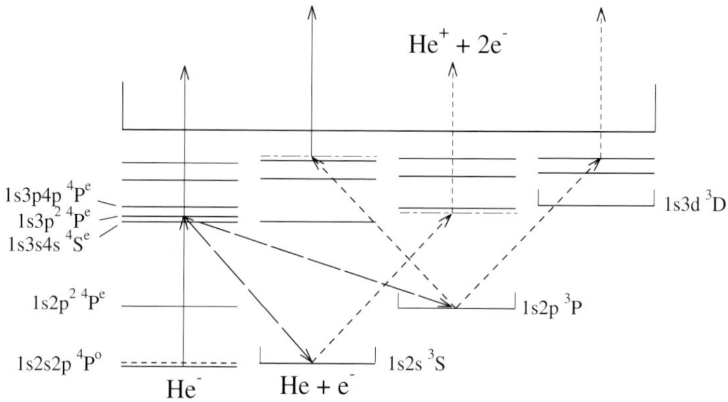

Figure 3. Level scheme for multiphoton single and double ionisation of He^-. Plain arrows: direct process; dashed arrows: sequential ionisation; long dashed arrows: spontaneous detachment.

In addition to this two-photon, sequential double ionisation, one has to consider the possibility of three-photon processes, provided these are resonant with some transitions in atomic helium. This requirement is fulfilled at specific wavelengths, where 1+1 REMPI of the photodetachment products occurs. With increasing photon energy, we find the $1s2p\ ^3P$ to $1s5s\ ^3S$, $1s2p\ ^3P$ to $1s5d\ ^3D$ and $1s2s\ ^3S$ to $1s3p\ ^3P$ transitions at 3.008, 3.079 and 3.188 eV respectively (figure 3). Their relative intensities are controlled by the branching ratio between He($1s2s\ ^3S$) and He($1s2p\ ^3P$) and by the associated oscillator strengths.

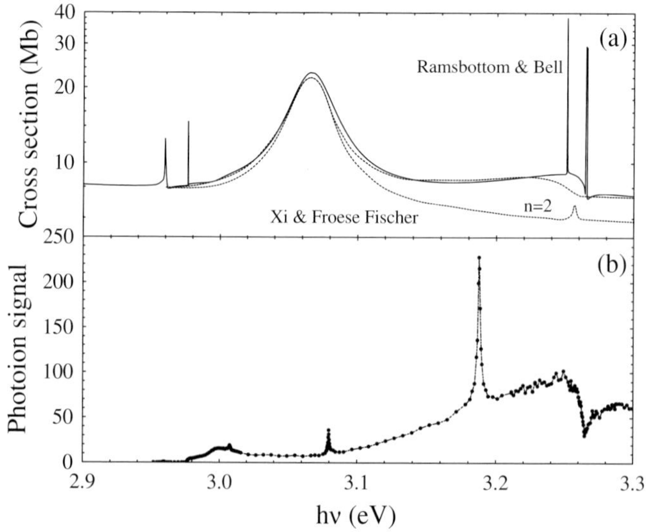

Figure 4. (a) Calculated photodetachment cross section. Dashed lines: Xi and Froese Fischer [8] - 'n=2'refers to calculations performed with a restricted basis (see text); full lines: Ramsbottom and Bell [9]. (b) Double ionisation signal measured with 1.5 mJ pulses (present work).

The background may be compared with the calculations of partial and total photodetachment cross sections calculated by Xi and Froese Fischer [8] and more recently by Ramsbottom and Bell [9] (figure 4). From the discussion of the sequential process given above, we may expect the He$^+$ signal to mimic the partial photodetachment cross section into $1s3l\ ^3L$ states. A tentative substraction of the calculation performed by Xi and Froese Fischer with the inclusion of $n = 2$ states only from the total cross section they obtained with a more complete basis produces fair agreement with the present results. The shoulder appearing at 2.985 eV does not seem to be predicted by any of the existing calculations, although the partial cross section for detachment into He($1s3l\ ^3L$) products was not specifically computed yet.

Among the three He$^-$ resonances listed in that wavelength range, namely the $1s3s4s\ ^4S$, $1s3p^2\ ^4P$ and $1s3p4p\ ^4P$ states located at 2.959, 3.072 and 3.265 eV respectively [3, 4], only the $1s3p4p\ ^4P$ resonance gives access to the He($1s3s\ ^3S$)+e$^-$ continuum. The resulting interference pattern is a marked minimum in the double ionisation yield. The decay of the $1s3p^2\ ^4P^e$ resonance into He($1s3s\ ^3S$)+e$^-$(εp) would produce the wrong parity and is therefore not contributing to the signal, while the $1s3s4s\ ^4S$ lies below threshold.

All structures listed so far clearly arise from sequential photon absorption and electron ejection. This may be expected in view of the relatively short lifetimes of the resonant states involved. The measured and calculated widths are 0.20, 50 and 0.61 meV for the $1s3s4s\ ^4S$, $1s3p^2\ ^4P$ and $1s3p4p\ ^4P$ states respectively [3, 4], which amounts to 20.7 ps, 82.7 fs and 6.8 ps for their respective lifetimes. These figures are all shorter than the interaction time (~90 ps), leaving room for sequential processes to take place. One notices however that the lifetime of the $1s3s4s\ ^4S$ state is of comparable duration (~20 ps). We shall dwell on this state in the remaining of the paper.

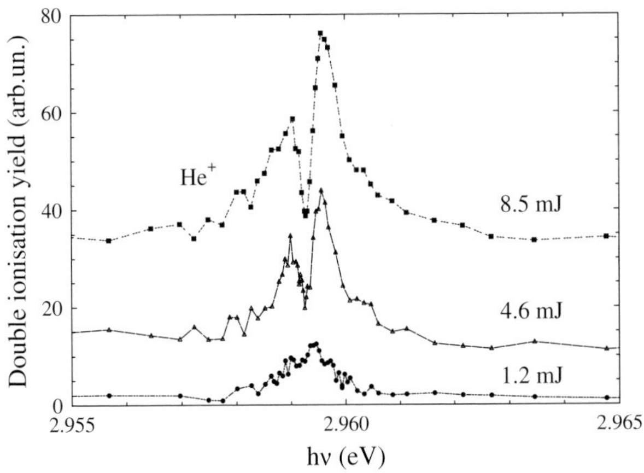

Figure 5. Double ionisation signal in the vicinity of the $1s3s4s\ ^4S$ state, as a function of laser pulse energy. The data are presented on a consistent scale.

As already visible on the spectrum presented in figure 2 on logarithmic scale, some double-ionisation signal is present below the He($1s3s\ ^3S$)+e$^-$ threshold at the approximate location of the $^4S^e$ resonance. A close-up view (figure 5) reveals a complicated peak-and-dip structure, with the dip coinciding exactly with the resonance position measured by Klinkmüller *et al.* [3]. An intensity-dependent study shows the development of the dip structure as the intensity is gradually increased to saturate the process.

Remembering that the two-photon, sequential double ionisation is not possible at that wavelength, one is left with the following interpretation. The signal may only arise from direct, double ejection of the two electrons, in a 1+1 REMPDI process, standing for resonance-enhanced multiphoton double ionisation. The background is attributed to three-photon, sequential processes in which single photon detachment, resonant or not, is followed by the two-photon ionisation of He($1s2s\ ^3S$) or He($1s2p\ ^3P$) products (figure 3). This multiphoton character is clearly demonstrated by the rapid rise of the background signal as the energy is increased.

It is very tempting to associate the observed peak-and-dip structure to the competition between REMPDI and resonant photodetachement, which are due to the population of one and the same level (figure 3). The latter process starts as soon as some population is transferred and may already occur when the ion is still in the wing of the gaussian laser profile, while the former obviously requires a stronger laser field, which is present in the middle of the laser beam. By the time this intensity is reached, spontaneous decay has already taken place, and depleted the two-level system. This depletion starts above the critical intensity needed for a π pulse, which explains why no dip is observed at low intensities.

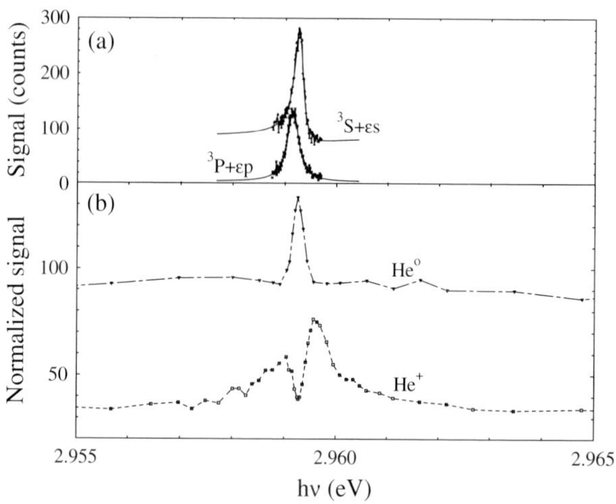

Figure 6. Comparison between photodetachment and two-photon double ionisation spectra. (a) Partial cross section measured by Klinkmüller *et al.* [3, 4]. (b) Photodetachment and two-photon double ionisation signal (present work).

An obvious verification of this interpretation would be to measure the photodetachment signal under the same experimental conditions. This measurement was performed in the pulsed-beam mode described above. The result shown in figure 6 agrees well with the measurement of Klinkmüller *et al.*regarding the position and the width of the resonance. Please note that we record total photodetachment while Klinkmüller *et al.* measured partial contributions in two separate experiments [3, 4]. The peak in the photodetachment signal nicely fits the dip observed in the two-photon double ionisation signal, supporting our interpretation.

4. Conclusions and perspectives

In a rather simple experiment, we have identified all possible pathways leading to the double ionisation of He^- by absorption of two or more photons. The detection of sequential ionisation itself brings new information about the very rich structure of this highly correlated system. Detailed calculations of the photodetachement process are still missing to assess the quality of our measurements, which appear to be complementary to those of Klinkmüller *et al.* [3, 4] as they are sensitive to n=3 products instead of the n=2 states probed in earlier work. The identification of a direct route towards double ionisation constitutes a more challenging issue, and our preliminary results are encouraging. One may speculate that the dip structure will nearly disappear as the interaction time is made shorter than the resonance lifetime. This would require to operate with a faster He^- beam and/or a smaller laser waist, and these measurements are in preparation. Clearly, a picosecond laser pulse would be ideally suited to investigate further the REMPDI process. Note that femtosecond laser pulses may not be the correct answer, as their broadband character would strengthen the role of non-resonant processes. Of course, the ultimate experiment requires the detection of both electrons in coincidence, with full determination of their ejection pattern, both angular and energetic. Such measurements are within reach of the new reaction microscopes equipped with time- and position-sensitive detectors [14, 15].

Acknowledgments

The authors wish to thank P J M van der Burgt for support in the early design of the present experiment, and T J Morgan for having suggested it. This work has been supported by the Belgian National Fund for Scientific Research (FNRS).

References

[1] Peterson J R, Bae Y K and Huestis D L 1985 *Phys. Rev. Lett.* **55** 692

[2] Walter C W, Seifert J A and Peterson J R 1994 *Phys. Rev.* A **50** 2257

[3] Klinkmüller A E, Haeffler G, Hanstorp D, Kiyan I Yu, Berzinsh U, Ingram C W, Pegg D J and Peterson J R 1997 *Phys. Rev.* A **56** 2788

[4] Klinkmüller A E, Haeffler G, Hanstorp D, Kiyan I Yu, Berzinsh U and Pegg D J 1998 *J. Phys. B: At. Mol. Opt. Phys.* **31** 2549

[5] Hanstorp D 1995 *Nucl. Instrum. Methods* B **100** 165

[6] Petrunin V V, Jacobsen M H, Madsen L B, Aseyev S A and Andersen T 2003 **90** 013002

[7] Berrah N, Bozek J D, Turri G, Akerman G, Rude B, Zhou H L and Manson S T 2002 *Phys. Rev. Lett.* **88** 093001

[8] Xi J and Froese Fischer C 1996 *Phys. Rev.* A **53** 3169

[9] Ramsbottom C A and Bell K L 1999 *J. Phys. B: At. Mol. Opt. Phys.* **32** 1315

[10] Brandefelt N and Lindroth E 2002 *Phys. Rev.* A **65** 032503

[11] Kristensen P, Pedersen U V, Petrunin V V, Andersen T and Chung K T 1997 *Phys. Rev.* A **55** 978

[12] Pedersen U V, Hyde M, Møller S P and Andersen T 2001 *Phys. Rev.* A **64** 012503

[13] Zeman H D 1977 *Rev. Sci. Instrum.* **48** 1079

[14] Pedersen H B, Jensen M J, Safvan C P, Urbain X and Andersen L H 1999 *Rev. Sci. Instrum.* **70** 3289

[15] Dorner R, Feagin J M, Cocke C L, Brauning H, Jagutzki O, Jung M, Kanter E P, Khemliche H, Kravis S, Mergel V, Prior M H, Schmidt-Böcking H, Spielberger L, Ullrich J, Unversagt M and Vogt T 1996 *Phys. Rev.* A **77** 1024

Inst. Phys. Conf. Ser. No. 183
Paper presented at Int. Conf. Electron and Photon Impact Ionization and Related Topics, Louvain-la-Neuve, 2004

Multiple photoionization of atoms and ions using the time-dependent close-coupling method

U Kleiman†, J Colgan‡, M S Pindzola† and F Robicheaux†

† Auburn University, Department of Physics, 206 Allison Laboratory, Auburn, AL 36849, USA

‡ Los Alamos National Laboratory, Theoretical Division T-4, Mail Stop B212, Los Alamos, NM 87545, USA

Abstract. Numerical calculations of partial cross sections for photoionization with excitation and total cross sections for double photoionization of the two-electron systems He and Li^+ have been performed employing the time-dependent close-coupling method. The calculations include one-photon ionization processes from both the ground state $1s^2\ ^1S$ and the two lowest metastable excited states $1s2s\ ^{1,3}S$. For the ground state $1s^22s\ ^2S$ of the three-electron system Li, partial cross sections for double photoionization with excitation and total cross sections for triple photoionization are presented.

1. Introduction

One-photon ionization processes in atoms and ions, which result in the ejection of two or even more electrons, are sensitive probes of electron-electron correlations. Although the pioneering works [1] go back to the late 1960s, it was not until the last decade that double and multiple photoionization processes have become a subject of intensive work by both theoreticians and experimentalists, as they have profited much from advanced computing facilities and third generation synchrotron radiation sources, as well as improved detection techniques.

The most basic of all these processes is the one-photon double ionization of the simplest two-electron system, the He atom. Consequently, this full fragmentation process has been thoroughly investigated where both total and differential cross sections have been considered. Although differential cross sections provide more insight into electron-electron correlations (e.g. Ref. [2]), the focus of this paper is on total cross sections. For the one-photon double ionization of He atoms from the ground state $1s^2\ ^1S$ and the two lowest metastable excited states $1s2s\ ^{1,3}S$ relative and, more challenging, absolute total cross sections have been calculated using different theoretical methods, like the many-body perturbation theory [3], the two screened Coulomb method [4], which has been later abandoned, the hyperspherical *R*-matrix method with semiclassical outgoing waves [5], the convergent close-coupling method [6, 7, 8, 9], the B-spline based R-matrix method [10, 11] and the time-dependent close-coupling method [12, 13, 14, 15, 16]. Comprehensive experimental studies of relative and absolute total cross sections for the one-photon double ionization of He atoms from the ground state, but not from excited states, have been performed by a few groups who employed different detection techniques, for example, time-of-flight spectroscopy [17, 18], a cylindrical mirror analyzer [19] and the very promising cold-target recoil-ion momentum spectroscopy [20].

Results for the one-photon double ionization of other two-electron systems of the He isoelectronic sequence are comparatively rare so far, which is primarily due to difficulties in handling these ions experimentally (e.g. Ref. [7] and references therein). For Li^+ ions, which are of interest here because comprehensive experimental studies of both the ground state $1s^2\ ^1S$ and metastable excited states are afoot [21], solely theoretical calculations have been carried out. Total cross sections have been obtained for the ground state [7, 11, 22] and for the excited state 1s2s ^1S [11, 23]. The excited state 1s2s ^3S has been considered just in the limit of high photon energies [23].

Closely related to one-photon double ionization, and likewise impossible in the absence of electron-electron correlations, is single photoionization with excitation, i.e. just one electron is ejected and the residual ion is left in an excited state. Partial cross sections for the ground state of He undergoing single photoionization with excitation have been obtained theoretically (e.g. Refs. [7, 9, 12, 24, 25]) and experimentally (e.g. Refs. [19, 26]). For the excited states 1s2s $^{1,3}S$ of He atoms, however, solely theoretical partial cross sections are available to date [8, 10, 15, 27]; as outlined above the same applies to the one-photon double ionization. Single photoionization with excitation of Li^+ ions has been considered even less. Partial cross sections have been evaluated for the ground state, only [7]. An extensive study involving not only the ground state but also the excited states 1s2s $^{1,3}S$ of Li^+ ions has been recently undertaken employing the time-dependent close-coupling method [28].

A step further towards a better understanding of electron-electron correlations is one-photon triple ionization and double ionization (with excitation) of three-electron systems, of which Li atoms are the simplest. Nevertheless, they are complex enough to exhibit intrashell as well as intershell correlations. In addition, Li has just one triple-photoionization threshold and the fragmentation into four particles is possible only by simultaneous ejection of all three electrons. Total cross sections for one-photon double ionization of Li have been calculated in the limit of high photon energies [29] and measured [30] while double ionization with excitation has been recently treated theoretically [31]. For the one-photon triple ionization of Li calculations [29, 31, 32] and measurements [33, 34] of total cross sections are available.

In this paper we first review our calculations of total and partial cross sections for one-photon double ionization and single photoionization with excitation from the ground state and the two metastable excited states 1s2s $^{1,3}S$ of He atoms employing the time-dependent close-coupling method, and then investigate the same processes in Li^+ ions. After that we move on to Li atoms, for which total and partial cross sections for one-photon triple ionization and double ionization with excitation from the ground state $1s^2 2s\ ^2S$ are presented.

2. Two-electron systems

Before discussing theoretical results for the two-electron systems He and Li^+ it is reasonable to outline the time-dependent close-coupling method which we use to describe one-photon double ionization and single photoionization with excitation. For details, the reader is referred to Ref. [12].

2.1. Theory

The time-dependent wavefunction for atomic or ionic two-electron systems with an initial ^{1,3}S symmetry exposed to dipole radiation may be split up into two parts

$$\Psi_{\text{total}}(\vec{r}_1,\vec{r}_2,t) = \Phi_0^{1,3\text{S}}(\vec{r}_1,\vec{r}_2)\,\mathrm{e}^{-\mathrm{i}E_0 t} + \Psi^{1,3\text{P}}(\vec{r}_1,\vec{r}_2,t). \tag{1}$$

As shown below, after some sufficient time $t = T_\text{s}$ has passed, total and partial cross sections can be obtained from the time-dependent wavefunction $\Psi^{1,3\text{P}}(\vec{r}_1,\vec{r}_2,t)$, which vanishes at $t = 0$.

The wavefunction $\Phi_0^{1,3\text{S}}(\vec{r}_1,\vec{r}_2)$ for the initial state (ground state 1s^2 ^{1}S or excited states 1s2s ^{1,3}S) is obtained by an expansion in coupled spherical harmonics

$$\Phi_0^{1,3\text{S}}(\vec{r}_1,\vec{r}_2,\tau) = \sum_l \frac{P_{ll}^{1,3\text{S}}(r_1,r_2,\tau)}{r_1 r_2}\, Y_{ll}^{1,3\text{S}}(\hat{r}_1,\hat{r}_2) \tag{2}$$

and subsequent relaxation of the time-dependent Schrödinger equation

$$-\frac{\partial\Phi_0^{1,3\text{S}}(\vec{r}_1,\vec{r}_2,\tau)}{\partial\tau} = H_{\text{target}}\,\Phi_0^{1,3\text{S}}(\vec{r}_1,\vec{r}_2,\tau) \tag{3}$$

in imaginary time ($\tau = \mathrm{i}t$), where the non-relativistic Hamiltonian is given by

$$H_{\text{target}} = -\frac{1}{2}\Delta_1 - \frac{Z}{r_1} - \frac{1}{2}\Delta_2 - \frac{Z}{r_2} + \frac{1}{|\vec{r}_1 - \vec{r}_2|} \tag{4}$$

with nuclear charge $Z = 2$ for He atoms and $Z = 3$ for Li$^+$ ions. Substituting expansion (2) into equation (3) results in a set of coupled partial differential equations for the two-electron radial wavefunctions $P_{ll}^{1,3\text{S}}(r_1,r_2,\tau)$ which is solved on a lattice (see sections 2.2 and 2.3) applying low-order finite difference methods [12] and beginning with $P_{ll}^{1\text{S}}(r_1,r_2,0) = P_{1\text{s}}(r_1)\,P_{1\text{s}}(r_2)\,\delta_{l,0}$ for the ground state and with $P_{ll}^{1,3\text{S}}(r_1,r_2,0) = \sqrt{1/2}\,[P_{1\text{s}}(r_1)\,P_{2\text{s}}(r_2) \pm P_{2\text{s}}(r_1)\,P_{1\text{s}}(r_2)]\,\delta_{l,0}$ for the excited states. Here, $P_{n\text{s}}(r)$ denotes bound electron radial wavefunctions of He$^+$ and Li^{2+}, respectively. The relaxation is carried out until the wavefunctions and energies are converged on the lattice. The energies obtained are typically within 1–2% of the binding energies of the atoms and ions, respectivley. Note, that the relaxation of the metastable excited state 1s2s ^{1}S requires a Schmidt orthogonalization, otherwise it would lead to the ground state.

The final state wavefunction $\Psi_{\text{total}}(\vec{r}_1,\vec{r}_2,t)$ describing the system after photoabsorption, satisfies the time-dependent Schrödinger equation which, in the weak-field limit where one-photon absorption is the only relevant process, reduces to

$$\mathrm{i}\frac{\partial\Psi^{1,3\text{P}}(\vec{r}_1,\vec{r}_2,t)}{\partial t} = H_{\text{target}}\,\Psi^{1,3\text{P}}(\vec{r}_1,\vec{r}_2,t) + H_{\text{field}}\,\Phi_0^{1,3\text{S}}(\vec{r}_1,\vec{r}_2)\,\mathrm{e}^{-\mathrm{i}E_0 t}. \tag{5}$$

Here, the classical electric field of amplitude $E(t)$ and frequency ω is assumed to be fully linearly polarized along the z-axis. In the length gauge its Hamiltonian may be written as

$$H_{\text{field}} = E(t)\,(r_1\cos\theta_1 + r_2\cos\theta_2)\,\sin(\omega t). \tag{6}$$

The time-dependent amplitude is ramped on linearly from zero to one over a quarter of the total time T_s necessary to time propagate equation (5) on a lattice for achieving convergence (see sections 2.2 and 2.3). For this, the wavefunction $\Psi^{^{1,3}\mathrm{P}}(\vec{r}_1,\vec{r}_2,t)$ is expanded in coupled spherical harmonics

$$\Psi^{^{1,3}\mathrm{P}}(\vec{r}_1,\vec{r}_2,t) = \sum_{l_1,l_2} \frac{P_{l_1l_2}^{^{1,3}\mathrm{P}}(r_1,r_2,t)}{r_1 r_2} Y_{l_1l_2}^{^{1,3}\mathrm{P}}(\hat{r}_1,\hat{r}_2) \tag{7}$$

with the triangle condition $\Delta(l_1,l_2,1)$, so that equation (5) along with (6) are transformed into a set of coupled partial differential equations for the two-electron radial wavefunctions $P_{l_1l_2}^{^{1,3}\mathrm{P}}(r_1,r_2,t)$ which, prior to time propagation, are zero for all angular momentum pairs l_1, l_2.

As mentioned above, different cross sections may be extracted from the wavefunction $\Psi^{^{1,3}\mathrm{P}}(\vec{r}_1,\vec{r}_2,t)$ after some sufficient time T_{s} has passed to ensure convergence. Employing standard projection techniques, the total cross section for one-photon double ionization and partial cross sections for single photoionization with excitation to a bound state nl may be expressed as follows

$$\sigma_{\mathrm{d}}^{^{1,3}\mathrm{P}} = \frac{\omega}{I} \frac{\partial \mathscr{P}_{\mathrm{d}}^{^{1,3}\mathrm{P}}}{\partial t}\bigg|_{t=T_{\mathrm{s}}} \qquad \sigma_{nl}^{^{1,3}\mathrm{P}} = \frac{\omega}{I} \frac{\partial \mathscr{P}_{nl}^{^{1,3}\mathrm{P}}}{\partial t}\bigg|_{t=T_{\mathrm{s}}} \tag{8}$$

where I denotes the intensity of the electric field. The corresponding probabilities are given by

$$\mathscr{P}_{\mathrm{d}}^{^{1,3}\mathrm{P}} = \sum_{l_1,l_2} \frac{2}{\pi}\int_0^\infty \mathrm{d}k_1 \frac{2}{\pi}\int_0^\infty \mathrm{d}k_2 \left| \int_0^\infty \mathrm{d}r_1 \int_0^\infty \mathrm{d}r_2\, P_{k_1l_1}(r_1)\, P_{k_2l_2}(r_2)\, P_{l_1l_2}^{^{1,3}\mathrm{P}}(r_1,r_2,t)\right|^2 \tag{9}$$

$$\begin{aligned}\mathscr{P}_{nl}^{^{1,3}\mathrm{P}} = \sum_{l_1,l_2} \Bigg[& \frac{2}{\pi}\int_0^\infty \mathrm{d}k_2 \left| \int_0^\infty \mathrm{d}r_1 \int_0^\infty \mathrm{d}r_2\, P_{nl}(r_1)\, P_{k_2l_2}(r_2)\, P_{l_1l_2}^{^{1,3}\mathrm{P}}(r_1,r_2,t)\right|^2 \\ & + \frac{2}{\pi}\int_0^\infty \mathrm{d}k_1 \left| \int_0^\infty \mathrm{d}r_1 \int_0^\infty \mathrm{d}r_2\, P_{k_1l_1}(r_1)\, P_{nl}(r_2)\, P_{l_1l_2}^{^{1,3}\mathrm{P}}(r_1,r_2,t)\right|^2 \Bigg]\end{aligned} \tag{10}$$

where $P_{nl}(r)$ and $P_{kl}(r)$ are bound and continuum radial wavefunctions of He^+, respectively. Their calculation is explained, for example, in Ref. [13]. Alternatively, and as applied to Li^+, projections can also be made onto products of bound states instead of distorted continuum states. The corresponding formulae are given in Refs. [12, 28].

2.2. *He atoms*

For He atoms equations (3) and (5) have been solved on a lattice consisting of 600×600 points with a mesh spacing of $\Delta r = 0.1$ a.u. where, for the ground state, angular momenta $l \leq 3$ and $l_1,l_2 \leq 4$ have been used in expansions (2) and (7), respectively, to achieve convergence. Angular momenta $l \leq 8$ and $l_1,l_2 \leq 9$ were needed for the excited state 1s2s ^{1}S while $l \leq 6$ and $l_1,l_2 \leq 7$ were sufficient for the other state 1s2s ^{3}S. Equation (5) has been time propagated for between 10 and 15 periods of the electric field where the propagation time had to be the longer for the smaller photon energy.

Figures 1 and 2 display total cross sections for one-photon double ionization and partial cross sections for single photoionization with excitation leaving the remaining bound electron in a shell with principal quantum number $n \leq 3$. Results are shown for ionization processes from the ground state and the two lowest metastable excited states 1s2s ^{1,3}S of He. For the latter states the first value calculated is located 10 eV above the respective double ionization

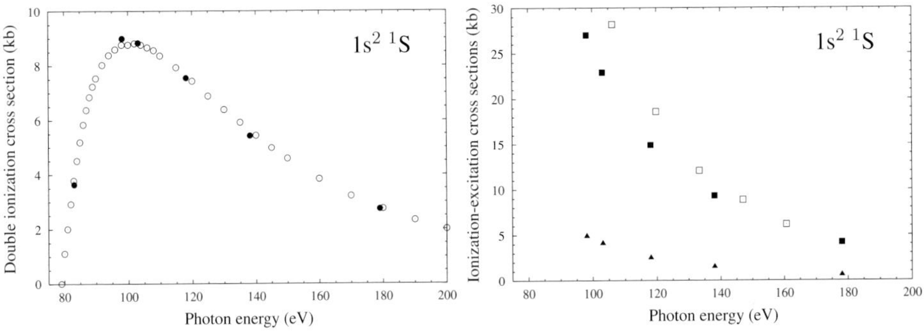

Figure 1. Double photoionization cross section $\sigma_d^{^1P}$ (•) and cross sections $\sigma_2^{^1P}$ (■) and $\sigma_3^{^1P}$ (▲) for single photoionization with excitation of He ($1s^2$, ^1S). The TDCC data are compared to experimental data for $\sigma_d^{^1P}$ (○) by Samson *et al* [18] and close-coupling data for $\sigma_2^{^1P}$ (□) by Jacobs and Burke [24].

thresholds. Concerning the ground state, we note that our theoretical values for the total double ionization cross section lie very close to the experimental curve of Ref. [18]. An earlier close-coupling calculation of the partial cross section for single ionization with excitation to the $n = 2$ shell [24] exhibits comparatively large deviations from our calculation, in particular at smaller photon energies. This is likely because of the small number of channels used in their close-coupling expansion due to scarce computational resources at that time. An inclusion of more channels may improve the agreement with our results. Turning to the excited singlet and triplet states of He, our results match the converged close-coupling results [8] very well, small deviations occur only in the partial cross section for single ionization from the 1s2s ^1S state with excitation to the $n = 3$ shell. The partial cross sections $\sigma_3^{^{1,3}P}$ evaluated in Ref. [27] deviate considerably from our results at smaller photon energies which might be due to the omission of the 1s, 2s and 2p channels in the close-coupling expansion. Further partial cross sections for single ionization with excitation to the $n = 4, 5$ shells are given in Ref. [15]. These partial cross sections are in the kb range and considerably smaller than those displayed in Figure 2. Thus, photoabsorption of He (1s2s, ^1S) most likely results in $n = 2$ or $n = 3$ states of He^+ and photoabsorption of He (1s2s, ^3S) by far yields the $n = 2$ state while double ionization is much less probable (note that the ordinate axes of Figure 2 are scaled in kb = 10^{-21} cm^2 and in Mb = 10^{-18} cm^2). In other words, the electron absorbing the photon more likely transfers a small amount of energy to the other electron (single photoionization with excitation) than a large one (double photoionization). As the electron-electron correlations in the metastable 1s2s ^3S state are weaker than in the metastable 1s2s ^1S state, one-photon double ionization from the excited triplet state shows a significant smaller total cross section than the excited singlet state. This difference is largest around the peaks of the total double ionization cross sections and decreases with increasing photon energy.

2.3. Li^+ ions

For Li^+ ions a lattice with 640×640 points and mesh spacing $\Delta r = 0.1$ a.u. has been chosen to solve equations (3) and (5). For the ground state $1s^2\ ^1S$ angular momenta $l \le 8$ and $l_1, l_2 \le 9$ have been used in expansions (2) and (7), respectively, whereas for the metastable excited states 1s2s $^{1,3}S$ angular momenta $l \le 5$ and $l_1, l_2 \le 6$ have been used. The large number

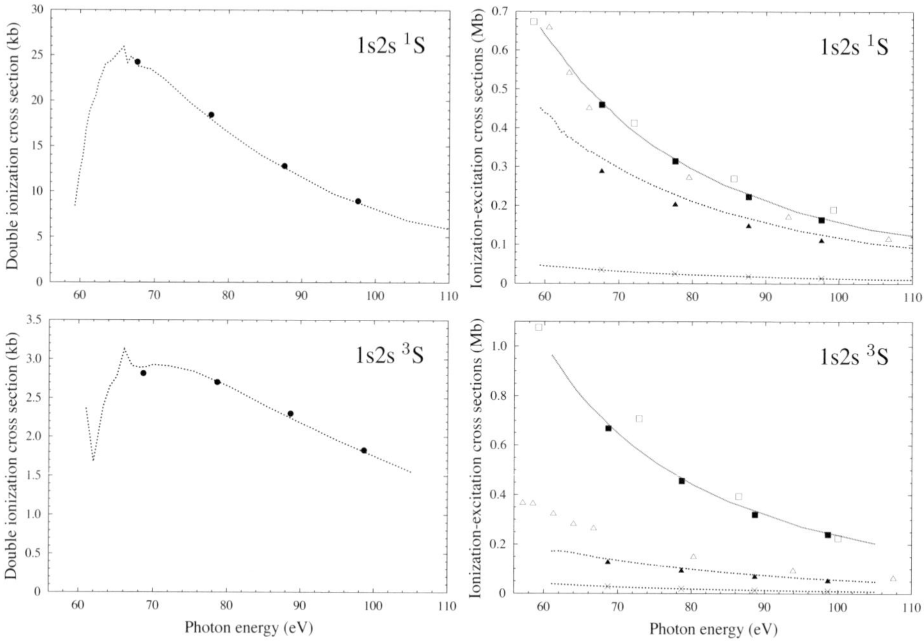

Figure 2. Double photoionization cross section $\sigma_d^{1,3P}$ (•) and cross sections $\sigma_1^{1,3P}$ (×), $\sigma_2^{1,3P}$ (■) and $\sigma_3^{1,3P}$ (▲) for single photoionization with excitation of He (1s2s, ^{1,3}S). The TDCC data are compared to convergent close-coupling data (lines) by Kheifets *et al* [8] and close-coupling data for $\sigma_2^{1,3P}$ (□) and $\sigma_3^{1,3P}$ (△) by Jacobs and Burke [27].

of angular momenta used for the ground state was needed to obtain convergent results for the partial cross sections for single ionization with excitation, especially below the double ionization threshold which is about 60 eV higher compared to the excited states. We note, that for He the corresponding energy difference is just about 20 eV and no partial cross sections have been calculated for photon energies below the double ionization thresholds discussed above. Again, mainly because of the partial cross sections for single ionization with excitation below the double ionization threshold, equation (5) has been time propagated for 25 periods of the electric field.

In Figures 3 and 4 we present cross sections for one-photon double ionization and single photoionization with excitation leaving the remaining bound electron of the Li^{2+} ion in a shell with $n \leq 3$. As for isoelectronic He, we show theoretical results for the ground state $1s^2\ ^1S$ and the two lowest metastable excited states 1s2s ^{1,3}S of Li^+. Further results, especially for photoionization with excitation to subshells will be discussed in Ref. [28]. For the ground state, absolute total cross sections for double photoionization have been obtained employing the convergent close-coupling method [7]. The overall agreement is good, in particular for photon energies below the peak. Without exception, for photon energies above the peak our values are slightly smaller than those of Ref. [7]. An experiment would be desirable to say more than that. For the excited singlet and triplet states of Li^+, to our knowledge, neither absolute theoretical nor absolute experimental cross sections are available so far. Our time-

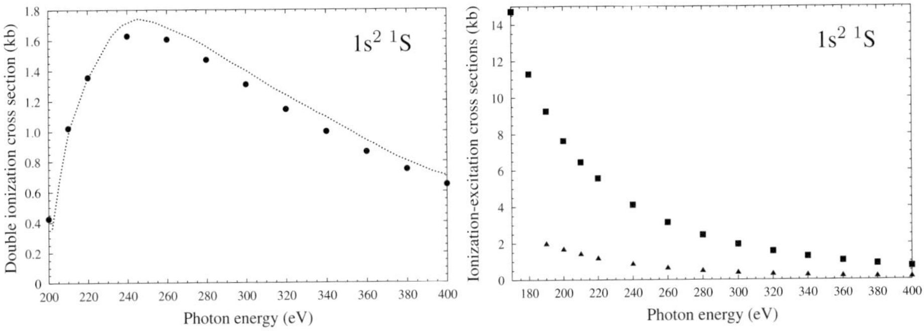

Figure 3. Double photoionization cross section $\sigma_d^{^1P}$ (•) and cross sections $\sigma_2^{^1P}$ (■) and $\sigma_3^{^1P}$ (▲) for single photoionization with excitation of Li^+ ($1s^2$, ^1S). The TDCC data are compared to convergent close-coupling data (– – –) by Kheifets and Bray [7].

dependent close-coupling calculation indicates that, for instance, photoabsorption processes from both the 1s2s ^1S and 1s2s ^3S states of Li^+ most likely results in Li^{2+} ions with the bound electron left in the $n = 2$ state. Note, that the cross sections for single ionization of the triplet state with excitation to $n = 1$ and $n = 3$ states are almost of the same magnitude. Double photoionization is much less probable. As for He, the total cross section for double ionization from the excited triplet state is the smaller one.

Comparing He atoms with Li^+ ions, we remark, that the respective total cross sections for one-photon double ionization differ considerably both in magnitude and in the position of the peak relative to the double photoionization threshold, whereas the partial cross sections for single photoionization with excitation are roughly of the same magnitude. The latter especially holds for the excited states.

3. Three-electron system: Li atoms

We now turn to a brief discussion of one-photon triple ionization and double photoionization with excitation of the open-shell atom Li, which is the simplest of all three-electron systems, further information can be found in Ref. [31].

As for the two-electron systems He and Li^+, the motion of the three electrons of Li is treated equally by time propagation of a nine-dimensional wavefunction according to the Schrödinger equation in the weak-field limit. The three radial dimensions are represented on a uniform lattice with 192^3 points and mesh spacing $\Delta r = 0.1$ a.u. while the six angular dimensions are represented by a coupled channel expansion. Angular momenta $l_i \leq 2$ have been used for the relaxation of the ground state $1s^22s$ ^2S. For the final state ^2P two different calculations have been carried out using $l_j \leq 2$ (spd) and $l_j \leq 3$ (spdf), respectively. In both cases, the time-dependent close-coupling equations have been propagated for 10 periods of the electric field. The probabilities for triple ionization and double ionization with excitation have been obtained by projecting the nine-dimensional final state wavefunction onto fully antisymmetric spatial and spin functions.

The left part of Figure 5 displays the total cross section for triple photoionization of the ground state of Li, where the first data point is about 60 eV above the triple ionization

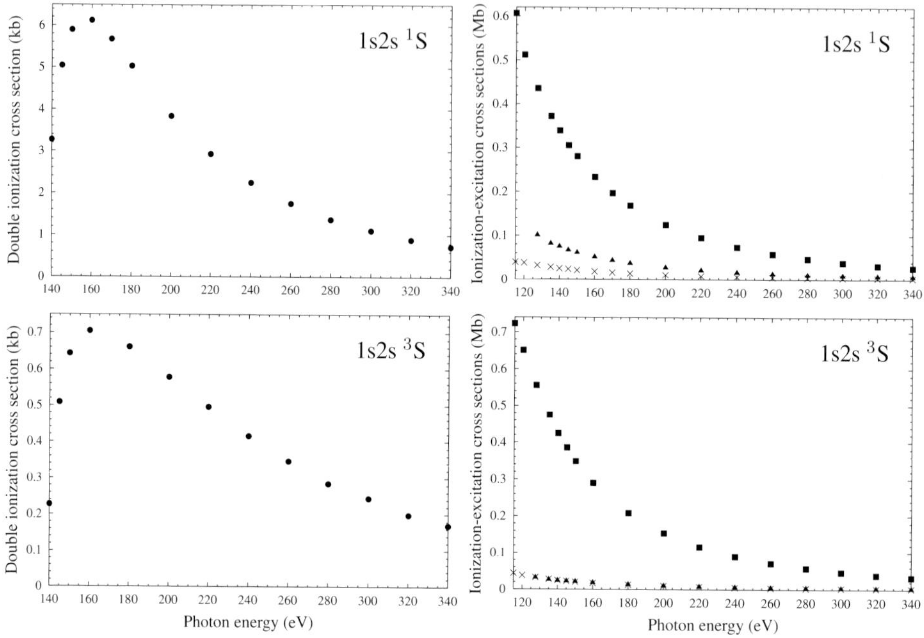

Figure 4. Double photoionization cross section $\sigma_d^{1,3P}$ (•) and cross sections $\sigma_1^{1,3P}$ (×), $\sigma_2^{1,3P}$ (■) and $\sigma_3^{1,3P}$ (▲) for single photoionization with excitation of Li^+ (1s2s, ^{1,3}S).

threshold. The calculation with $l_j \leq 2$ (spd) yields significantly smaller cross sections than the other calculation with $l_j \leq 3$ (spdf). That is, the triple photoionization cross section is difficult to converge, chiefly because of its very small quantity (b = 10^{-24} cm^2). The use of a larger mesh may improve the convergence but the present computational resources do not allow for this. Nonetheless, our calculations are supported by experimental values [33], which generally fall between our two data sets and which themselves show large error bars.

In the right part of Figure 5 we present partial cross sections for double photoionization with excitation leaving the remaining bound electron of Li^{2+} in one of the subshells 1s, 2s and 2p. In the energy region considered, the bound electron is as likely to be left in the subshell 1s as in 2s. Thus, the 'shake-off' process, in which the dominant mechanism is the double ionization of the 1s^2 core of Li followed by the ionization of the remaining 2s electron via 'shake-off' for triple ionization [33, 34], cannot be the only mechanism that contributes to double (triple) ionization. An alternative mechanism may be based on an 'internal scattering' process, where a 1s or 2s electron is initially ionized, which then collides with the remaining two electrons and ionizes one (two) electron(s) for double (triple) ionization.

4. Conclusions

In this paper we have reviewed our time-dependent close-coupling calculations of total and partial cross sections for double photoionization and single photoionization with excitation

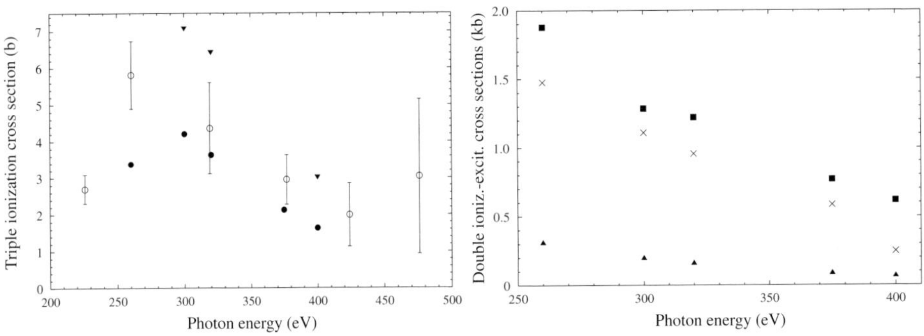

Figure 5. Triple photoionization cross sections σ_t (•) (spd) and σ_t (▼) (spdf) and cross sections σ_{1s} (×), σ_{2s} (■) and σ_{2p} (▲) for double photoionization with excitation of Li ($1s^2 2s$, ^2S). The TDCC data are compared to experimental data (○) by Wehlitz *et al* [33].

of He atoms and have extended our investigation to isoelectronic Li^+ ions. As for He, for which not only total cross sections, but also differential cross sections have been calculated [13, 14, 15, 16], we plan to also obtain differential cross sections for the double photoionization of Li^+, since they provide further insight into electron-electron correlations. For the more challenging three-electron atom Li, recently calculated total and partial cross sections for triple photoionization and double photoionization with excitation have been presented. An extension of this method to get differential cross sections for Li is planned.

Acknowledgments

This work was supported by grants from the U.S. Department of Energy and the National Science Foundation. The bulk of the computational work was carried out at the Center for Computational Sciences at Oak Ridge National Laboratory and the National Energy Research Scientific Computing Center at Lawrence Berkeley National Laboratory. The computational work was further supported by the National Science Foundation cooperative agreement ACI-9619020 through computing resources provided by the National Partnership for Advanced Computational Infrastructure at the San Diego Supercomputer Center.

References

[1] Carlson T A 1967 *Phys. Rev.* **156** 142–9
Byron F W and Joachain C J 1967 *Phys. Rev.* **164** 1–9

[2] Briggs J S and Schmidt V 2000 *J. Phys. B: At. Mol. Opt. Phys.* **33** R1–48
King G C and Avaldi L 2000 *J. Phys. B: At. Mol. Opt. Phys.* **33** R215–84

[3] Hino K, Ishihara T, Shimizu F, Toshima N and McGuire J H 1993 *Phys. Rev.* **48** 1271–6
Pan C and Kelly H P 1995 *J. Phys. B: At. Mol. Opt. Phys.* **28** 5001–12

[4] Proulx D and Shakeshaft R 1993 *Phys. Rev.* A **48** R875–8
Pont M and Shakeshaft R 1995 *J. Phys. B: At. Mol. Opt. Phys.* **28** L571–7

[5] Selles P, Malegat L and Kazansky A K 2002 *Phys. Rev.* A **65** 032711

[6] Kheifets A S and Bray I 1996 *Phys. Rev.* A **54** R995–7; 2000 *Phys. Rev.* A **62** 065402

[7] Kheifets A S and Bray I 1998 *Phys. Rev.* A **58** 4501–11

[8] Kheifets A S, Ipatov A, Arifin M and Bray I 2000 *Phys. Rev.* A **62** 052724

[9] Kheifets A S 2001 *J. Phys. B: At. Mol. Opt. Phys.* **34** L247–52

[10] van der Hart H W, Meyer K W and Greene C H 1998 *Phys. Rev.* A **57** 3641–5

[11] van der Hart H W and Feng L 2001 *J. Phys. B: At. Mol. Opt. Phys.* **34** L601–9

[12] Pindzola M S and Robicheaux F 1998 *Phys. Rev.* A **57** 318–24; 1998 *Phys. Rev.* A **58** 779–80

[13] Colgan J, Pindzola M S and Robicheaux F 2001 *J. Phys. B: At. Mol. Opt. Phys.* **34** L457–66

[14] Colgan J and Pindzola M S 2002 *Phys. Rev.* A **65** 032729

[15] Colgan J and Pindzola M S 2003 *Phys. Rev.* A **67** 012711

[16] Colgan J and Pindzola M S 2004 *J. Phys. B: At. Mol. Opt. Phys.* **37** 1153–64

[17] Wehlitz R, Heiser F, Hemmers O, Langer B, Menzel A and Becker U 1991 *Phys. Rev. Lett.* **67** 3764–7
Levin J C, Armen G B and Sellin I A 1996 *Phys. Rev. Lett.* **76** 1220–3

[18] Samson J A R, Stolte W C, He Z-X, Cutler J N, Lu Y and Bartlett R J 1998 *Phys. Rev.* A **57** 1906–11

[19] Bizau J M and Wuilleumier F J 1995 *J. Electron Spectrosc. Relat. Phenom.* **71** 205–24

[20] Dörner R *et al* 1996 *Phys. Rev. Lett.* **76** 2654–7

[21] Landers A 2004 *Private communication*

[22] Dalgarno A and Sadeghpour H R 1992 *Phys. Rev.* A **46** R3591–3
Kornberg M A and Miraglia J E 1994 *Phys. Rev.* A **49** 5120–3

[23] Forrey R C, Sadeghpour H R, Baker J D, Morgan J D and Dalgarno A 1995 *Phys. Rev.* A **51** 2112–6

[24] Jacobs V L and Burke P G 1972 *J. Phys. B: At. Mol. Phys.* **5** L67–70

[25] Berrington K A, Burke P G, Fon W C and Taylor K T 1980 *J. Phys. B: At. Mol. Phys.* **15** L603–8
Chang T N 1980 *J. Phys. B: At. Mol. Phys.* **13** L551–7
Gorczyca T W and Badnell N R 1997 *J. Phys. B: At. Mol. Opt. Phys.* **30** 3897–911

[26] Wehlitz R *et al* 1997 *J. Phys. B: At. Mol. Opt. Phys.* **30** L51–8

[27] Jacobs V L and Burke P G 1972 *J. Phys. B: At. Mol. Phys.* **5** 2272–81

[28] Kleiman U, Pindzola M S and Robicheaux F 2004 *Phys. Rev.* A submitted

[29] van der Hart H W and Greene C H 1998 *Phys. Rev. Lett.* **81** 4333–6

[30] Huang M-T, Wehlitz R, Azuma Y, Pibida L, Sellin I A, Cooper J W, Koide M, Ishijima H and Nagata T 19 99 *Phys. Rev.* A **59** 3397–401

[31] Colgan J, Pindzola M S and Robicheaux F 2004 *Phys. Rev. Lett.* **93** 053201

[32] Pattard T and Burgdörfer J 2001 *Phys. Rev.* A **63** 020701

[33] Wehlitz R, Huang M-T, DePaola B D, Levin J C, Sellin I A, Nagata T, Cooper J W and Azuma Y 1998 *Phys. Rev. Lett.* **81** 1813–6

[34] Wehlitz R, Pattard T, Huang M-T, Sellin I A, Burgdörfer J and Azuma Y 2000 *Phys. Rev.* A **61** 030704

Inst. Phys. Conf. Ser. No. 183
Paper presented at Int. Conf. Electron and Photon Impact Ionization and Related Topics, Louvain-la-Neuve, 2004

Ab initio Double Photoionization Cross Sections For Helium Down To Threshold

P. Selles†, L. Malegat†, and A. K. Kazansky‡

†Laboratoire d'Interaction du Rayonnement X Avec la Matière (UMR 8624), Université Paris-Sud, 91405 Orsay, France

‡Fock Institute of Physics, The University of St Petersburg, St Petersburg, 198504 Russia

Abstract. A complete set of absolute cross sections, going from the total, fully integrated to the fully differential ones, is calculated for the double photoionization (DPI) of helium at 100 meV above threshold. The reported *ab-initio* calculations, obtained from the hyperspherical $\mathscr{R}$-matrix with semiclassical outgoing waves (H$\mathscr{R}$M-SOW) method [2, 3], are confronted successfully with predictions based on the famous Wannier threshold model for double escape, together with absolute differential measurements [1] available at this energy. As a result, the H$\mathscr{R}$M-SOW method, which has already been validated in the intermediate 20-60 eV energy range [4, 5, 6, 7, 8], can be definitively considered as an achieved theory of the PDI of helium.

1. Introduction

1.1. Present state of helium PDI

The description of the three-body Coulomb problem that can not be accounted for in an independent-particle approach is a long standing difficulty in atomic physics. Here, we focus on the prototypal process of double photoionization (DPI) of helium by one photon. It has known a revival of interest in the last ten years because of decisive progress both in experimental and theoretical activities. On the experimental side, considerable technical advances in the third generation synchrotron radiation sources and in the multidetection techniques have allowed the production of accurate information, especially of triply differential cross sections (TDCS), in a large variety of dynamical situations. The low and intermediate excess energy range, from 1 up to 60 eV, has been widely studied [9, 10, 11, 12, 13, 4, 5]. Absolute TDCS have even been measured at 0.1 eV above threshold [1]. On the theoretical side, significant progress came from the development of three novel *ab-initio* numerical methods that treat the two electrons as genuinely equivalent and circumvent the difficulty of assessing an explicit asymptotic boundary condition. The time dependent close coupling (TDCC) method [14, 15], the exterior complex scaling (ECS) method [16] applied only recently to PDI of helium [17], and the H$\mathscr{R}$M-SOW method [2, 3, 7] have however been applied so far to excess energies of a few eV to a few tens of eV above the threshold, excluding the very vicinity of this latter.

1.2. *The close vicinity of the threshold: a real challenge to* ab-initio *theories*

Surely the study of the very close vicinity of the PDI threshold is the most stringent test to evaluate the accuracy of *ab-initio* theories faced with the full complexity of the three body Coulomb problem. In return, two main drastic conditions must be fulfilled to obtain accurate double escape cross sections:

(i) Because of the long range Coulomb interactions, the method should be devised to take into account a very large part of the double continuum configuration space. Namely, as emphasized in the third section, at 0.1 eV above threshold, the asymptotic region in which the electron-electron repulsion can be ignored is only reached for electron-nucleus distances about at least 3 10^3 a.u., while the redistribution of the fluxes between single and double escape continues up to 10^6 a.u.

(ii) The three-body channel must be disentangled from competitive single continuum two-body channels, the number of which can reach at least one hundred at 0.1 eV above threshold. The method should be able to extract information contained in 0.015% of the total signal. Effectively, the double escape integrated cross section (DEICS) can be evaluated to 0.09 kb according to the Wannier-type threshold law experimentally established by Kossmann *et al* [18], as the total single escape integrated cross section can be interpolated to about 730 kb from the compilation by Bizau and Wuilleumier [19].

In this work H$\mathscr{R}$M-SOW calculations are presented at 100meV above threshold. They provide for the first time a complete set of *ab-initio* theoretical cross sections in the very close vicinity of the threshold. The second section is devoted to a presentation of the main tools and principles of the method and to its implementation at threshold. The third section displays the whole set of results. Single differential cross sections (SDCS) enlighten the ability of the method to face (i) and (ii) requirements listed above. The difficult problem of the extraction of the DEICS mentioned in (ii) is solved. TDCS are shown in comparison with renormalized absolute measurements from Huetz and Mazeau [1]. Doubly differential cross sections (DDCS), and more particularly asymmetry parameters β, are discussed. In the last section predictions based on Wannier's ideas [20], which remain so far the reference in the discussion of near threshold double photoionization, are checked against the H$\mathscr{R}$M-SOW *ab-initio* calculations.

2. H$\mathscr{R}$M-SOW method at threshold

2.1. *Main principles*

Only the main principles and tools of the H$\mathscr{R}$M-SOW method are introduced in this section. More detailed information can be found in [2, 3]. The method relies on the following inhomogeneous stationary equation:

$$\left[\mathscr{H}_0-(E_0+\omega)\right]\Phi_1(\vec{r}_1,\vec{r}_2)=-\frac{1}{2}\vec{\mathscr{E}}_0\cdot\vec{D}_G\Psi_0(\vec{r}_1,\vec{r}_2) \tag{1}$$

where Ψ_0 and E_0 are the initial He ground state and its energy, Φ_1 is the photoabsorption wave function, ω the frequency of the incident light, $\vec{\mathscr{E}}_0$ the amplitude of the electric field vector, and

$\vec{D}_G$ the dipole operator that can be expressed in different gauges. The photoabsorption wave function Φ_1 contains contributions from the entire set of ionization channels, including the single-ionization with and without excitation, as well as the double-ionization channel. It is expanded over bipolar harmonics and hyperradial functions depending on the hyperspherical coordinates $R = \sqrt{r_1^2 + r_2^2}$ and $\alpha = \arctan(r_1/r_2)$. The solution is then obtained by splitting the configuration space into two regions depending on the value of R. In the inner region $R < R_0$ a full quantum treatment is used. It is complemented with an outgoing adiabatic wave condition imposed on the hypersphere $R = R_0$. As a result the solution can be extracted on the border of the inner region. It is then propagated throughout the outer region assuming a semiclassical behaviour of its R dependency. The unique ability of the H$\mathscr{R}$M-SOW method to provide the photoabsorption wave function up to distances that clearly belong to the asymptotic region allows a direct evaluation of the outgoing fluxes, giving access to the cross sections from their very definition.

2.2. *Implementation of the method at threshold*

This method is well suited to face the (i) and (ii) requirements described in the introduction.

(i) Only one variable, R, goes to ∞, so that the computational task associated to the propagation up to the asymptotic region is significantly simplified. The propagation from R_0=60 up to R_∞=10^6 a.u. is performed after 190000 propagation steps. It is light enough to be completed on a plain PC (Pentium III 450 MHz, 768 Mb RAM).

(ii) When R gets large enough, single-ionization becomes confined within two intervals of width $\Delta\alpha = r_{n_{\max}}/R$ around $\alpha = 0$ and $\pi/2$, where $r_{n_{\max}}$ measures the extension of the highest $He^+(n = n_{\max})$ Rydberg state that can be excited significantly by the the photon absorption. Valuable physical information regarding the dynamics of this complex multichannel photoionization process is obtained thanks to a performant description of the radial correlation coordinate α. It is discretized on a 4000 point variable stepsize mesh with a much higher density of representative points in the critical $\alpha = 0$ and $\alpha = \pi/2$ regions.

The increase of the wavelength associated to each of the two escaping electrons is however a difficulty when approaching the threshold. Indeed a full quantum treatment of the problem is required up to distances where the relative variation of the potential is negligible on an interval the extension of which is comparable to the electronic wavelength. Such a distance should be at least 100 a.u. at 0.1 eV. The most achieved version of our method [7] allows an extension of the inner region up to only R_0=60 a.u. The first adiabatic channels can nevertheless be identified to the $He^+(n$=1,2,3) singly ionized states and subtracted from the wavefunction before its propagation. This subtraction validates somewhat the semiclassical model for the R-propagation because it allocates to the local momentum a meaningful effective charge $\simeq 5$, close to the Wannier one. Even if the requirement of an inner region of at least one hundred a.u. is not strictly fulfilled, the semiclassical approximation we impose prematurely can be checked against the reliability of the absolute integrated cross sections provided by the method. These values will be displayed in the next section.

3. Results

3.1. SDCS

Evolution with R of the renormalized fluxes that converge at infinity to the SDCS is followed in Fig. (1). We give these fluxes the same name as their respective asymptotic limits, although it is clearly an abuse of language. Only the end of the propagation, going from R=10^3 to R=10^6 a.u., is displayed. As R increases, the middle part of this pattern widens, flattens, and goes down, while the two side peaks get much narrower. In other words, the flux is redistributed between single and double ionization channels in the course of the system expansion up to distances as large as 10^6 a.u. Contrary to higher energies where the side peaks become rapidly so narrow

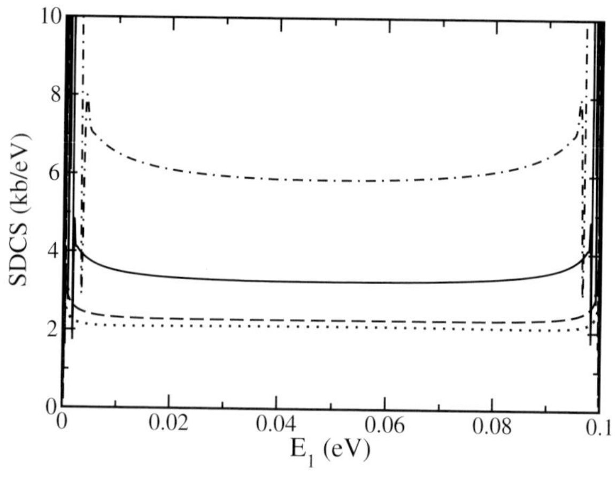

Figure 1. SDCS from H$\mathcal{R}$M-SOW calculations plotted in kb/eV versus the energy of the measured electron in eV. Dashed-dotted line: $R = 10^3$ a.u. Full line: $R = 10^4$ a.u. Dashed line: $R = 10^5$ a.u. Dotted line: $R = 10^6$ a.u.

that they can hardly be distinguished from the vertical axis (Fig. (8) of [3]), the transition between the two regimes is visible up to R=10^6 a.u. at 0.1 eV. This is clearly an indication that numerous He^+ Rydberg states can be excited significantly by the photon absorption.

Fig. (2) is an enlargement of SDCSs in the small α region. On the right the plateau corresponds to the double escape events whereas the oscillations on the left are related to the single escape events. The vertical lines locate the $He^+(n_{\max})$ excited state. At R=3 10^3 (R=10^4) a.u., the $\Delta\alpha$ confinement can be evaluated to 0.165 rad (0.125 rad), which gives a spatial extent $r_{n_{\max}}$ of about 500 (1250) a.u. and leads to an evaluation of $n_{\max} \simeq 20$ ($n_{\max} \simeq 35$) at R=3 10^3 (R=10^4) a.u. respectively. So the study of SDCS clearly indicates that higher and higher He^+ Rydberg states are populated throughout the propagation in the outer region. At R=10^6 a.u., the $\Delta\alpha$ confinement could be evaluated to 2.5 10^{-2} rad, and $n_{\max}$ to at least one hundred. Similarly, higher and higher He^+ Rydberg states are populated as the excess energy above threshold is decreased. A similar study of the SDCS at $4eV$ above threshold would indicate a $n_{\max}$ value of about 12 only at R=10^4 a.u. These evolutions result from electron-electron correlations, the importance of which has already been illustrated through the analysis of TDCS [21]. These latter have been considered so far as the quantities containing the most detailed information on

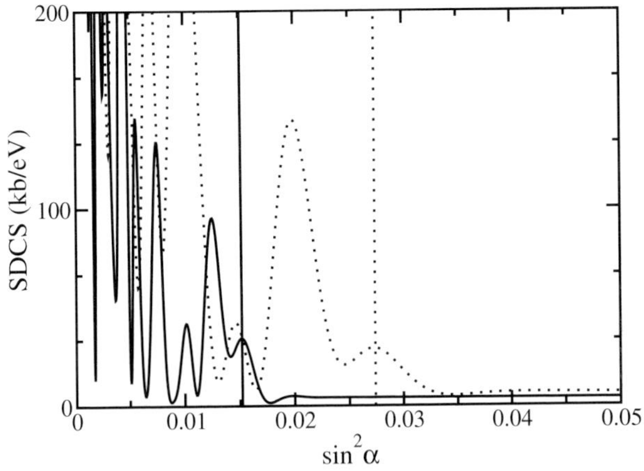

Figure 2. SDCS from H$\mathscr{R}$M-SOW calculations plotted versus $\sin^2\alpha$ on an expanded small α scale. The vertical lines locate the separation between the double and single escape regimes. Dotted line: R=3 10^3 a.u. Full line: R=10^4 a.u.

the dynamics of the process. This section clearly demonstrates that valuable complementary information can also been obtained from the investigation of SDCSs.

3.2. Integrated cross sections

In order to provide a complete check of the consistency of the method at threshold, the DEICS is firstly obtained by integrating the SDCS over the energy of the first electron. Performing this integration requires to have at one's disposal a converged and reliable SDCS free of any contamination by single escape events.

Fig. (3) displays on an expanded vertical scale the evolution of SDCSs' shapes obtained at R=10^6 a.u. with respect to parameters of the H$\mathscr{R}$M-SOW calculations. Clearly, a N=4000 point α grid, together with an expansion of the photoabsorption wave function on L=50 partial waves in the outer region are required to provide accurate and converged SDCS. As a comparison, only 20 partial waves were needed at 20 eV above threshold. The far expansion of electron-electron correlations at threshold is obviously responsible for such an effect. The suppression of the single-ionization events is realized thanks to an even sixth-order polynomial parametrization, in the same manner than at higher energies [3]. The result of this parametrization is shown on Fig. (3). It can only been distinguished from the calculated SDCS in regions where the energy of one of the electrons is less than 5 10^{-3} eV. This value defines the limit beyond which the calculated differential cross sections can be considered free of contamination by the single-ionization channels. Because it cannot be associated to any sound singly excited state, its physical meaning is however questionable. Even if the DEICS obtained by integrating the polynomial approximation gives the reliable value of 0.106 kb, we decided to implement another method, by subtracting the cross sections for single-ionization to the ground and excited ionic states from the total ionization cross section, so that:

$$\sigma^{++} = \sigma_{photo} - \sum_{n=1}^{n=3} \sigma^{+}_{adia}(n) - \sum_{n=4}^{n=24} \sigma^{+}_{proj}(n) - \sum_{n=25}^{n=100} \sigma^{+}_{fit}(n) \qquad (2)$$

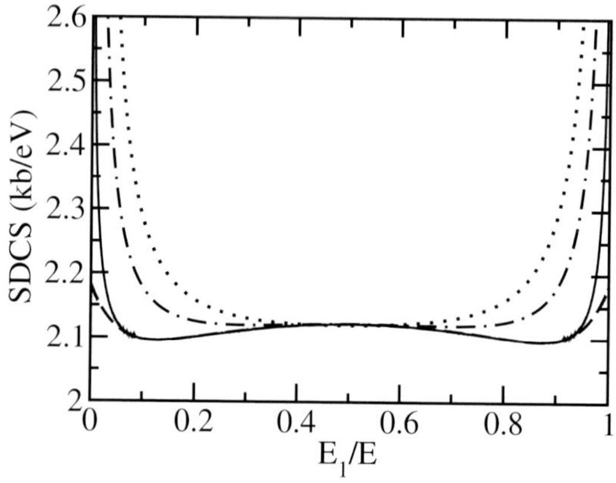

Figure 3. SDCS from H$\mathscr{R}$M-SOW calculations at R=10^6 a.u. plotted in kb/eV versus the relative energy of the measured electron. Dotted line: N=1500, L=25. Dashed-dotted line: N=1500, L=40. Full line: N=4000, L=50. Dashed line: polynomial parametrization.

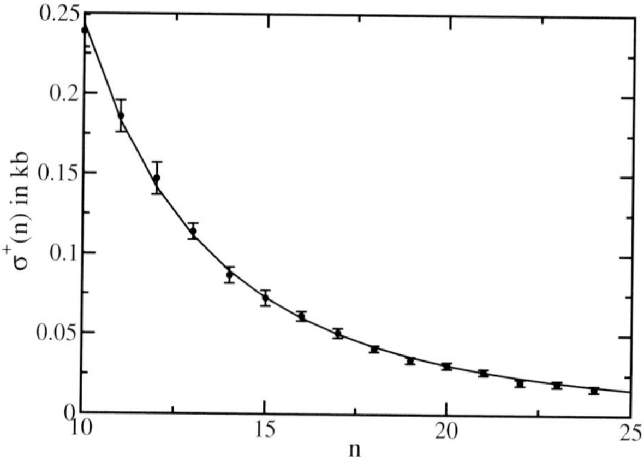

Figure 4. Partial single ionization cross sections for excited $He^+(n)$ states obtained by the projection technique. The $\sigma^+(n)$ are plotted in kb versus n. Points with error bars: H$\mathscr{R}$M-SOW calculations. Full line: fractional n^{-3} fit.

The obtention of the total photoabsorption cross section σ_{photo} and of the partial σ^+(n=1-3) cross sections corresponding to adiabatic channels is documented in [3]. The σ^+_{proj}(n=4-24) are evaluated by projecting the photoabsorption wave function Φ_1 onto the wave functions of the singly ion ized states n=4-24. The projection is defined using the r_1 or r_2 coordinates, but tractable at R fixed and large. On those grounds, it is validated if the useful part of the integral is entirely contained in $\Delta\alpha_{proj}$ sectors on the border of the $r_1^2 + r_2^2 = R^2$ sphere in the (r_1, r_2) plane. For this procedure to be meaningful, each n level has to be projected out at a convenient $R(n)$ that fulfills the condition $R(n) = \Delta r_n / \Delta\alpha_{proj}$ where Δr_n is the spatial extent of the n state. As a result the projection is made possible up to n=24 with a 4000

point α grid. To obtain the last n=25-100 contributions, an extrapolation procedure is used. The asymptotic evolution of the $\sigma^{+}_{proj}(n)$ from n=10 up to n=24 is fitted to a fractional n^{-3} law. Fig. (4) displays the result of this operation. An excellent chisquare of 0.5 is obtained. This n^{-3} dependency seems to indicate that a few ℓ levels are populated for each high lying n excited state. Such a conclusion deserves extensive research, in connection with the prediction by Fano [22], experimentally confirmed [23], that large angular momentum is acquired at threshold by each of the two electrons.
As a conclusion, the overall procedure leads to two determinations of the DEICS, namely 0.106 kb by the parametrization technique and 0.098 kb by the projection technique, both in excellent agreement with the experimental value 0.09 kb extracted from [18].

3.3. TDCS

In this section absolute TDCSs from H$\mathscr{R}$M-SOW calculations are compared with renormalized absolute experimental TDCSs. The justification of such a renormalization is explained firstly. The procedure used by Huetz and Mazeau [1] to obtain absolute TDCS is based on a relation between differential fivefold $N^{(5)}$ and integral $N^{(0)}$ number of counted e-e coincidences. The fivefold quantities are related to the variables E_1, Ω_1 and Ω_2, the energy and spherical angles of the escaping electrons. The $N^{(5)}$ number is obtained within the experimental acceptance bandwidth Δx_5. The procedure is free of any uncontrolled instrumental parameter thanks to the 4π solid angle detection efficiency of the experiment. The measured TDCS are approximate averaged fivefold quantities:

$$\bar{\sigma}^{(5)}_{exp} = \frac{N^{(5)}}{N^{(0)}} \frac{\sigma^{(0)}}{\Delta x_5} \qquad (3)$$

where $\sigma^{(0)}$ is nothing else than the DEICS, identified as σ^{++}. The calibration is performed thanks to the $\sigma^{(0)}$ value from Kossmann *et al.* This approximation leads to an underestimation of TDCS by a factor of 1.67. We have evaluated this latter with respect to the exact averaged fivefold $\bar{\sigma}^{(5)}$ that takes into account the photon beam energy distribution, calculates exactly the volume integral corresponding to the experimental bandwidths and uses as the scaling reference for $\sigma^{(0)}$ the H$\mathscr{R}$M-SOW estimation of 0.098 kb. The renormalized experimental TDCS are then plotted on Fig. (5) in comparison with H$\mathscr{R}$M-SOW calculations. They have been obtained under equal energy sharing between the two electrons. The agreement is excellent both in shapes and in magnitudes. This is all the more rewarding because quite zero signals are concerned. The bottom right panel gives in addition a quantitative information on the role played by electron-electron correlations. The full (dotted) line corresponds to H$\mathscr{R}$M-SOW calculations in which propagation has been performed up to R=3 10^3 a.u. with (without) electron-electron repulsion. Clearly electron-electron correlations confine the two electrons in two different half-planes. More interestingly, the dashed curve contains in fact two superimposed calculations coming from a propagation throughout the outer region up to R=10^6 a.u. with electron-electron repulsion and another one with this latter up to R=3 10^3 a.u. and without further on. The role of the correlations up to R=3 10^3 a.u. is certainly to determine then to fix the relative directions for the two electron escapes. The persisting evolution in the redistribution of the intensities between 3 10^3 and 10^6 a.u. indicates that correlations go on playing a role, even if the $1/r_{12}$ term has almost vanished during the propagation. They certainly influence the further evolution of the system via initial conditions imposed at 3 10^3 a.u. As a comparison, electron-electron repulsion has been proved to be negligible for electron-nucleus distances larger than 60 a.u. at excess energies above 10 eV [21].

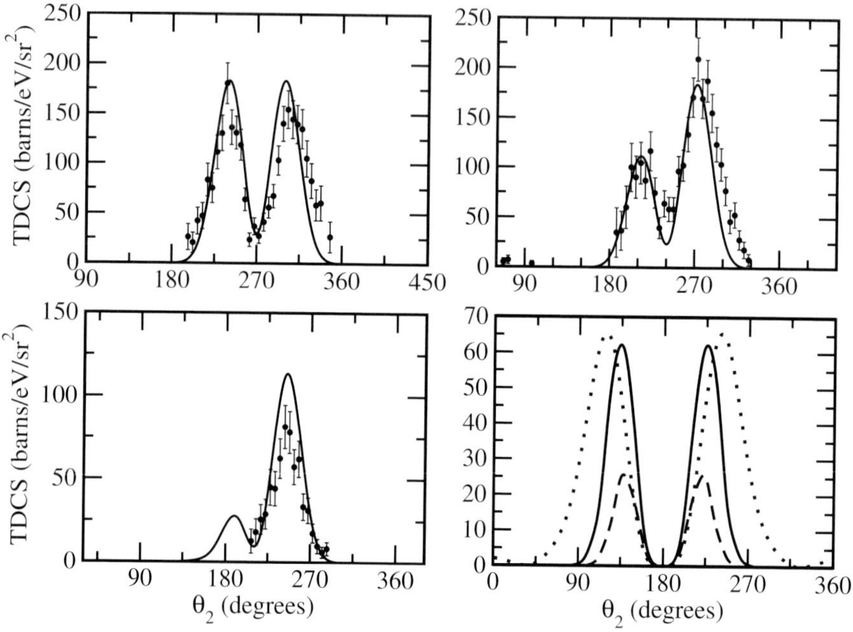

Figure 5. Coplanar TDCS in b(eV)$^{-1}$(sr)$^{-2}$ plotted versus the azimuthal angle θ_2 of the second electron in degrees, the first electron being detected at various but fixed θ_1 angle with respect to the electric field of the linearly polarized light. Top left panel: θ_1=90°. Top right panel: θ_1=60°. Bottom left panel: θ_1=30°. In these three panels TDCS are averaged according to the experimental conditions of Huetz and Mazeau [1]. Full line: H$\mathscr{R}$M-SOW calculations. Points with error bars: experimental data multiplied by a factor 1.67. Bottom right panel: H$\mathscr{R}$M-SOW TDCS for θ_1=0°. Full line: R=3 10^3 a.u. with e-e repulsion. Dotted line: R=3 10^3 a.u. without e-e repulsion. Dashed lines: R=10^6 a.u. with e-e repulsion.

Fig. (6) shows the evolution of coplanar TDCS at fixed excess energy E=0.1 eV with respect to the energy sharing. The first electron is detected along the electric field of the linearly polarized light. The most striking feature of the figure is its spectacular stability, all the more spectacular because it can be compared to the overturnings observed for the same complementary dynamics at 25 eV [7] and 40 eV [12] above threshold. The shapes are similar and the scales differ only by a factor 1.4. The kinematic node at θ_{12}=180° is slightly filled up for unequal energy sharing. All these observations indicate that the gerade amplitude is widely dominant, and quite energy sharing independent. Moreover the TDCS obtained by neglecting the ungerade amplitude and simulating the gerade one by a Gaussian shape of FWHM=47°, is in excellent agreement with the H$\mathscr{R}$M-SOW calculations. Similar studies at 0.15 eV and 0.05 eV above threshold lead to Gaussian shapes of FWHM=50° and 42° respectively for the gerade amplitudes.

3.4. DDCS

In this section we are interested in the β parameter that governs the behaviour of the DDCS. It depends *a-priori* on the sharing of the energy between the two electrons, and not only on

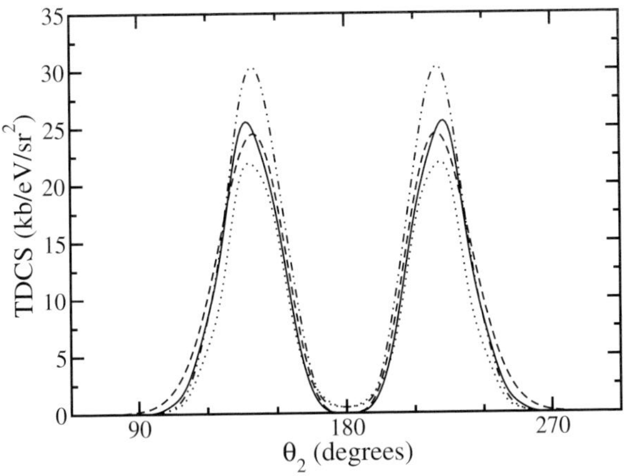

Figure 6. TDCS at R=10^6 a.u. in the same geometric conditions as in the bottom right panel of Fig. (5). Full line: H$\mathscr{R}$M-SOW E_1=E_2. Dashed line: Gaussian fit E_1=E_2. Dotted line: H$\mathscr{R}$M-SOW E_1/E_2=$1/4$. Dashed-dotted line:H$\mathscr{R}$M-SOW E_1/E_2=4.

the total energy available above threshold. The β parameters computed by the H$\mathscr{R}$M-SOW method under the assumption of full linear polarization of the incident light are displayed on Fig. (7). They are plotted for two fixed excess energies with respect to the energy sharing,

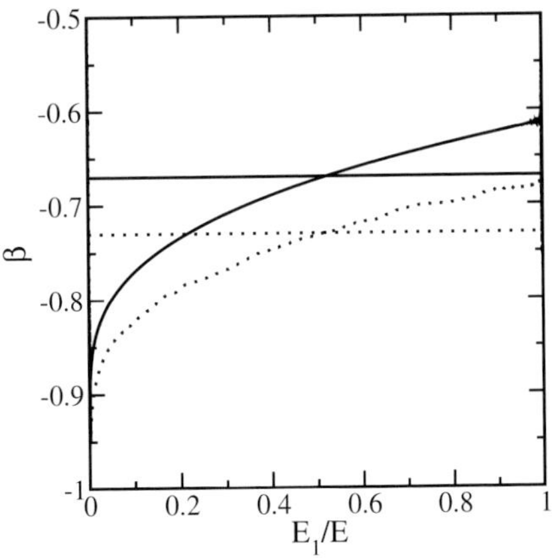

Figure 7. Asymmetry parameter β from H$\mathscr{R}$M-SOW calculations versus the energy ratio E_1/E for two different excess energies above threshold. Full lines: E=0.1 eV. Dotted line: E=0.05 eV. Horizontal lines: predictions from the Wannier model.

although they are usually studied for a fixed energy sharing with respect to the excess energy. It is the first time that such negative values approaching the -1 limit are produced. T his limit is reached when the measured electron has a quasi zero energy, in contradiction with theoretical calculations [24] and experimental results at 2 eV above threshold [25], according to which the strongest manifestation of electron correlation is found to occur for equal energy sharing. The figure exhibits an energy sharing dependency, while predictions coming from the Wannier model lead to energy sharing independent parameters. Neglecting the ungerade amplitude and simulating the gerade one by the Gaussian shape determined just above through the analysis of TDCS give two constant values -0.67 and -0.73 at respectively 0.15 and 0.05 eV above threshold. These values coming from the Wannier model correspond mostly to the H$\mathscr{R}$M-SOW ones calculated at equal energy sharing.

4. Wannier model

The first analysis of the double electronic escape at threhold was performed by Wannier in 1953 [20]. His idea was based on the description of the long range dynamics in the vicinity of the saddle point of the potential, using an ad hoc boundary condition to simulate the short range dynamics. The predictions made by Wannier have so far remained the reference in the discussion of near-threshold double photoionization. These predictions concern all usually measured cross sections from the fully integrated to the fully differential ones. Our H$\mathscr{R}$M-SOW calculations for TDCS are in agreement with Wannier predictions. The ungerade amplitude is demonstrated to be negligible and a Gaussian shape the FWHM of which is consistent with a $84^o E^{1/4}$ dependency is obtained for the gerade amplitude. The H$\mathscr{R}$M-SOW DDCS are slightly energy sharing dependent. The β parameter tends towards -1 only if E_1 and E jointly go to zero. A quick look at Fig. (1) shows a perfeclty flat SDCS. A detailed investigation of Fig. (2) however reveals a slight enhancement of SDCS in the vicinity of equal energy sharing. Because we have not yet undertaken a systematic study of the vicinity of the threshold we are unable to confirm the well known Wannier law giving a E^m dependency of the DEICS at threshold, with $m = 1.056$ for the DPI of helium. We have nevertheless obtained a reliable value of 0.098 kb, in excellent agreement with the 0.09 kb value extrapolated from the experimental $1.02E^{1.05}$ law established by Kossmann [18] in the two first eV above threshold.

5. Conclusion

We have displayed for the first time at 100 meV above threshold a wide selection of computed ab-initio ionization cross sections, ranging from the fully-differential to the fully-integrated type, and regarding single- as well as double-ionization processes. The reliability of these data, confronted successfully with predictions based on the famous Wannier threshold model of double escape, together with absolute differential measurements available at this energy, definitively validates our H$\mathscr{R}$M-SOW model as a complete theory for the PDI of helium. The distinctive features of the method provide an exact description of the electron correlations up to required electron-nucleus distances of at least $3\,10^3$ a.u. This property gives to our H$\mathscr{R}$M-SOW method a significant and distinctive advantage over TDCC and ECS models that are actually unable to develop an accurate solution up to so large distances. Moreover since the H$\mathscr{R}$M-SOW method allows one to follow the evolution of the photoabsorption wave function throughout its propagation in the outer region, it provides valuable physical information on the dynamics of the system expansion, and more particularly on the way the total flux is redistributed over the double and single escape channels.

To obtain a complete investigation of the threshold we should nevertheless study a whole range of excess energies from the very threshold up to a few eV above this latter. The hyperspherical time-dependent method with semiclassical outgoing waves (HTD-SOW) method [26], which is tightly linked to the initial H$\mathscr{R}$M-SOW method seems well suited to take up the threshold region challenge in the near future.

References

[1] Huetz A and Mazeau J 2000 *Phys. Rev. Lett.* **85** 530

[2] Malegat L, Selles P and Kazansky A K 2000 *Phys. Rev. Lett.* **85** 4450

[3] Selles P, Malegat L and Kazansky A K 2002 *Phys.Rev.A* **65** 032711

[4] Collins S A, Cvejanović S, Dawson C, Reddish T J, Seccombe D P, Huetz A, Malegat L, Selles P, Kazansky A K, Danjo A, Soejima K, Okuno K and Yagishita A 2002 *Phys.Rev.A* **65** 052717

[5] Seccombe D P, Collins S A, Reddish T J, Selles P, Malegat L, Kazansky A K and Huetz A 2002 *J. Phys. B: At. Mol. Phys.* **35** 3767

[6] Bolognesi P, Kheifets A S, Bray I, Malegat L, Selles P, Kazansky A K and Avaldi L 2003 *J. Phys. B: At. Mol. Phys.* **36** L1

[7] Selles P, Malegat L, Huetz A, Kazansky A K, Seccombe D P and Reddish T J 2004 *Phys.Rev.A* **69** 052707

[8] Bolognesi P, Zitnik M, Malegat L, Selles P, Turri G, Coreno M, Camilloni R and Avaldi L 2004 *J. Phys. B: At. Mol. Phys.* **37** 2285

[9] King G C and Avaldi L 2000 *J. Phys. B: At. Mol. Phys.* **33** R215

[10] Dawson C, Cvejanović S, Seccombe D P, Reddish T J, Maulbetsch F, Huetz A, Mazeau J and Kheifets A K 2000 *J. Phys. B: At. Mol. Phys.* **33** 265

[11] Collins S A, Huetz A, Reddish T J, Seccombe D P and Soejima K 2001 *Phys.Rev.A* **64** 062706

[12] Bolognesi P, Camilloni R, Coreno M, Turri G, Berakdar J, Kheifets A S and Avaldi L 2001 *J. Phys. B: At. Mol. Phys.* **34** 3193

[13] Achler M, Mergel V, Spielberger L, Dörner R, Azuma Y and Schmidt-Böcking H 2001 *J. Phys. B: At. Mol. Phys.* **34** 965

[14] Pindzola M S and Robicheaux F 1996 *Phys.Rev.A* **54** 2142

[15] J. Colgan, M.S. Pindzola and F. Robicheaux, J. Phys. B **34**, L457 (2001); J. Colgan and M.S. Pindzola, Phys.Rev.A **65** 032729 (2002)

[16] Rescigno T N, Baertschy M, Isaacs W A and McCurdy C W 199 *Nature* **286** 2474; Baertschy M, Rescigno T N, Isaacs W A, Li X and McCurdy C W 2001 *Phys.Rev.A* **63** 022712; McCurdy C W, Horner D A and Rescigno T N 2001 *Phys.Rev.A* **63** 022711

[17] McCurdy C W, Horner D A, Rescigno T N and Martin F 2004 *Phys.Rev.A* **69** 032707

[18] Kossmann H, Schmidt V and Andersen T 1988 *Phys.Rev.Lett.* **60** 1266

[19] Bizau J M and Wuilleumier F 1995 *J.Electron Spectrosc.Relat.Phenom.* **71** 205

[20] Wannier G H 1953 *Phys. Rev.* **90** 817

[21] Malegat L 2004 *Physica Scripta* **T110** 83

[22] Fano U 1974 *J. Phys. B: At. Mol. Phys.* **7** L401-4

[23] Hammond P, Read F H and King G C 1988 *J. Phys. B: At. Mol. Phys.* **21** 3123-3129

[24] Maulbetsch F and Briggs J S 1992 *Phys.Rev.Lett.* **68** 2004

[25] Wehlitz R, Heiser F, Hemmers O, Langer B, Menzel A and Becker U 1991 *Phys.Rev.Lett.* **67** 3764

[26] Kazansky A K, Selles P and Malegat L, 2003 *Phys.Rev.A* **68** 052701

Inst. Phys. Conf. Ser. No. 183
Paper presented at Int. Conf. Electron and Photon Impact Ionization and Related Topics, Louvain-la-Neuve, 2004

On the Green function for two charged particles in the presence of a Coulomb center of force

Robin Shakeshaft

Physics Department, University of Southern California, Los Angeles CA 90089-0484, USA

Abstract. Computationally useful expressions for the one-particle Coulomb Green function are developed for all energies, thereby facilitating the evaluation of the two-particle Green function as a standard convolution integral.

1. Introduction

The Green function for two charged particles moving under the influence of their mutual interaction and the interaction with a fixed Coulomb center of force can be expressed as a convolution of two-one particle Coulomb Green functions — see e.g. [1]. This convolution entails an integration over one-particle energies. In this paper we provide computationally useful expressions for the one-particle Coulomb Green function over all energies, with special treatment of the low- and high-energy limits. We present results of a numerical application to two-electron escape from the negative hydrogen ion by one-photon absorption.

2. One-Particle Coulomb Green Function

The radial Green function for a particle of charge $-e$ to move with energy $\mathscr{E}$ in the presence of a fixed center of charge Ze is

$$g_l(r,r';\mathscr{E}\pm i0) = -\frac{k}{\mathscr{E}}\mathscr{R}_{kl}(r_<)\mathscr{I}^{\pm}_{kl}(r_>) \tag{1}$$

where the wavenumber k is a two-valued function of the energy,

$$k = \sqrt{2\mu\mathscr{E}}/\hbar, \tag{2}$$

where $\mathscr{R}_{kl}(r)$ and $\mathscr{I}^{\pm}_{kl}(r)$ are the regular and irregular Coulomb wavefunctions, respectively, which have the asymptotic forms

$$\mathscr{R}_{kl}(r) \sim \sin\theta_{kl}(r), \quad r\sim\infty \tag{3}$$

$$\mathscr{I}^{\pm}_{kl}(r) \sim e^{\pm i\theta_{kl}(r)}, \quad r\sim\infty \tag{4}$$

where β is the dimensionless parameter

$$\beta = \frac{Z}{a_0 k}, \tag{5}$$

with a_0 the Bohr radius, and where

$$\begin{aligned} \theta_{kl}(r) &= kr - \beta \ln(2kr) - l\pi/2 + \sigma_{kl}, && (6) \\ \sigma_{kl} &= \arg\Gamma(l+1+i\beta). && (7) \end{aligned}$$

The "physical" branch of $g_l(r,r';\mathscr{E})$ is defined on the "physical" energy sheet $0 < \arg(\mathscr{E}) < 2\pi$, where it is bounded as r and/or r' increase. The "unphysical" branch is defined on the "unphysical" energy sheet $-2\pi < \arg(\mathscr{E}) < 0$, where it is *un*bounded as r and/or r' increase. The two sheets are joined along "unitarity cut", i.e. the positive energy axis.

2.1. Low-Energy Limit

2.1.1. Repulsive Coulomb Potential We first consider motion in a repulsive Coulomb potential, Ze^2/r with $Z > 0$. The energy $\mathscr{E}$ is negligible compared to the potential when r lies in the range

$$\sqrt{2Zr/a_0} \ll 2\beta \tag{8}$$

or, equivalently, when $r \ll r_0$ where r_0 is the classical turning point of a particle impinging on the Coulomb barrier with zero angular momentum:

$$r_0 = (2a_0/Z)\beta^2. \tag{9}$$

In this region $kr \ll \beta$ and the oscillations of the Coulomb wavefunctions are governed largely by the Coulomb potential; as a result the wavefunctions can be expanded in powers of $\mathscr{E}$ through the series representations — see e.g. formulae 14.4.1 and 14.4.2 of [2] —

$$\mathscr{R}_{kl}(r) = \frac{C_l(\beta)(2l+1)!}{(2\beta)^{2l+1}(kr)^l} \sum_{n=2l+1}^{\infty} a_n(\beta)(2Zr/a_0)^{n/2} I_n(2\sqrt{2Zr/a_0}) \tag{10}$$

and

$$\mathscr{I}^{\pm}_{kl}(r) \sim \frac{b_l(\beta)}{C_l(\beta)(2l+1)(kr)^l} \sum_{n=2l+1}^{\infty} (-1)^n a_n(\beta)(2Zr/a_0)^{n/2} K_n(2\sqrt{2Zr/a_0}) \tag{11}$$

where $I_n(z)$ and $K_n(z)$ are the regular and irregular modified Bessel functions, $C_l(\beta)$ is the Coulomb normalization factor, i.e.

$$C_l(\beta) = \sqrt{\frac{2\pi\beta}{e^{2\pi\beta} - 1}} \tag{12}$$

and

$$\begin{aligned} & a_{2l} = 0, \;\; a_{2l+1} = 1, \;\; a_{2l+2} = 0, && (13) \\ & 4\beta^2(n-2l)a_{n+1} = -na_{n-1} - a_{n-2}, \;\; n \geq 2l+2, && (14) \\ & b_l(\beta) \sum_{n=2l+1}^{\infty} (-1)^n (n-1)! a_n(\beta) \sim 2. && (15) \end{aligned}$$

The series representation of $\mathscr{R}_{kl}(r)$ converges for $r < r_0$, but not uniformly; $a_n(\beta)$ decreases in magnitude as $(1/2\beta)^n$ as n increases, and hence $a_n(\beta)(2Zr/a_0)^{n/2}$ is of order $(r/r_0)^{n/2}$. In contrast, the series representations of $\mathscr{I}^+_{kl}(r)$ and $\mathscr{I}^-_{kl}(r)$ are only asymptotically convergent. Due to the singular nature of $K_n(z)$ at the origin each term of the series on the right side of Eq. (11) contains the irregular contribution

$$(-1)^n a_n(\beta)(2Zr/a_0)^{n/2}K_n(2\sqrt{2Zr/a_0}) \approx (-1)^n a_n(\beta)\frac{(n-1)!}{2}, \quad r \approx 0, \tag{16}$$

and the sum of these contributions, $1/b_l(\beta)$, converges only asymptotically as $\beta \to 0$. Hence the inequality $r \ll r_0$ must be supplemented by $\beta \ll 1$ for Eq. (11) to be valid.

The probability for a particle to penetrate the repulsive Coulomb barrier decreases exponentially with increasing penetration. Conversely, the regular wavefunction for a particle inside the Coulomb barrier increases exponentially as the distance from the origin increases. Indeed, it follows from Eq. (10) and the large-z asymptotic behavior $I_n(z) \approx e^z/\sqrt{2\pi z}$ that $\mathscr{R}_{kl}(r)$ is proportional to $e^{2\sqrt{2Zr/a_0}}$ for $1 \ll \sqrt{2Zr/a_0} \ll 2\beta$. Since the oscillations of $\mathscr{R}_{kl}(r)$ in the region $r \ll r_0$ are highly damped compared to those in the region $r \gg r_0$ the series representation of $\mathscr{R}_{kl}(r)$ cannot be matched to the asymptotic form on the right side of Eq. (3). To say this another way, an undamped standing wave to the right of the classical turning point cannot be matched to an exponentially decreasing wavefunction to the left of the turning point. Equivalently, the series representations of $\mathscr{I}^+_{kl}(r)$ and $\mathscr{I}^-_{kl}(r)$ cannot distinguish betw een outgoing- and ingoing-wave behavior. In fact, the relative difference between $\mathscr{I}^+_{kl}(r)$ and $\mathscr{I}^-_{kl}(r)$ in the region $r \ll r_0$ is of order $e^{-2\pi\beta}$, as can be seen from

$$\mathscr{I}^+_{kl}(r) - \mathscr{I}^-_{kl}(r) = 2i\mathscr{R}_{kl}(r), \tag{17}$$

noting that $C_l^2(\beta) \approx 2\pi\beta e^{-2\pi\beta}$.

The resulting series representation of the radial Green function for a *repulsive* Coulomb potential, applicable when both $r_> \ll r_0$ and $\beta \ll 1$, is

$$\begin{aligned} g_l(r,r',\mathscr{E} \pm i0) \quad \sim \quad & -\frac{(2l)!b_l(\beta)}{Ze^2(2Zr_</a_0)^l(2Zr_>/a_0)^l} \\ & \times \left(\sum_{n=2l+1}^{\infty} a_n(\beta)(2Zr_</a_0)^{n/2}I_n(2\sqrt{2Zr_</a_0}) \right) \\ & \times \left(\sum_{n'=2l+1}^{\infty} (-1)^{n'}a_{n'}(\beta)(2Zr_>/a_0)^{n'/2}K_{n'}(2\sqrt{2Zr_>/a_0}) \right). \end{aligned} \tag{18}$$

Since $a_n(\beta)$ and $b_l(\beta)$ are analytic functions of $\mathscr{E}$ [it being understood that the series representation of $b_l(\beta)$ must be truncated at the point where the series is relatively insensitive to the inclusion of further terms] the right side of Eq. (18) is analytic in $\mathscr{E}$ and has no branch point at $\mathscr{E} = 0$. This is not surprising since the difference between the outgoing- and ingoing Coulomb wavefunctions, $\mathscr{I}^+_{kl}(r) - \mathscr{I}^-_{kl}(r)$, is negligible in the region $r \ll r_0$ if $e^{-2\pi\beta} \ll 1$. Suppose, however, that we allow $\mathscr{E}$ to become complex. As $|\arg\mathscr{E}|$ increases the factor $e^{-2\pi\beta}$ increases in magnitude, and becomes exponentially large as $\mathscr{E}$ moves from the sector $|\arg\mathscr{E}| < \pi$ to $\pi < |\arg\mathscr{E}| \leq 2\pi$; accordingly, the difference between $\mathscr{I}^+_{kl}(r)$ and $\mathscr{I}^-_{kl}(r)$ becomes transparent. In fact, the Coulomb wavefunctions are multivalued functions of $\mathscr{E}$. Let us change $\arg\mathscr{E}$

from 0 to $\pm 2\pi$; it follows from Eq. (4) that in the asymptotically large-r region the outgoing- and ingoing wavefunctions transform as

$$\mathscr{I}^{+}_{kl}(r) \rightarrow e^{\mp\pi\beta}\mathscr{I}^{-}_{kl}(r), \quad \arg\mathscr{E}: 0 \rightarrow \pm 2\pi, \tag{19}$$

$$\mathscr{I}^{-}_{kl}(r) \rightarrow e^{\pm\pi\beta}\mathscr{I}^{+}_{kl}(r), \quad \arg\mathscr{E}: 0 \rightarrow \pm 2\pi. \tag{20}$$

Since the Coulomb normalization factor transforms as $C_l(\beta) \approx \sqrt{2\pi\beta} e^{-\pi\beta} \rightarrow \sqrt{2\pi\beta}$ as $|\arg\mathscr{E}| \rightarrow 2\pi$, the right side of Eq. (11) decreases by a factor of order $e^{-\pi\beta}$. Evidently, the right side of Eq. (11) represents $\mathscr{I}^{+}_{kl}(r)$ when $\pi < \arg\mathscr{E} \leq 2\pi$ and $\mathscr{I}^{-}_{kl}(r)$ when $-\pi > \arg\mathscr{E} \geq -2\pi$. Hence the right side of Eq. (18) represents the analytic continuation of $g_l(r,r',\mathscr{E}+i0)$ and $g_l(r,r',\mathscr{E}-i0)$ as $\mathscr{E}$ moves into the regions $0 < \arg\mathscr{E} \leq 2\pi$ and $0 > \arg\mathscr{E} \geq -2\pi$, respectively.

2.1.2. Attractive Coulomb Potential Now we consider an attractive Coulomb potential, $-Ze^2/r$, again with $Z > 0$. To obtain the series representations of the Coulomb wavefunctions for an attractive potential we must let $Z \rightarrow -Z$, and $\beta \rightarrow -\beta$, on the right sides of Eqs. (10) and (11). The coefficients $a_n(\beta)$ and $b_n(\beta)$ are unaffected since they are even in β. Furthermore, since $(2Zr/a_0)^{n/2} I_n(2\sqrt{2Zr/a_0})$ is an analytic function of Z there is no ambiguity in letting Z change sign in the regular Coulomb wavefunction; we obtain

$$\mathscr{R}_{kl}(r) = \frac{C_l(\beta)(2l+1)!}{(2\beta)^{2l+1}(kr)^l} \sum_{n=2l+1}^{\infty} (-1)^n a_n(\beta)(2Zr/a_0)^{n/2} J_n(2\sqrt{2Zr/a_0}) \tag{21}$$

where $J_n(z)$ is the regular Bessel function and where now (with β defined as before) the Coulomb normalization factor is

$$C_l(\beta) = \sqrt{\frac{2\pi\beta}{1-e^{-2\pi\beta}}}. \tag{22}$$

However, $(2Zr/a_0)^{n/2} K_n(2\sqrt{2Zr/a_0})$ has a logarithmic singularity at the origin and therefore is a multivalued function of Z. Consequently, the particular branch of the irregular wavefunction, $\mathscr{I}^{+}_{kl}(r)$ or $\mathscr{I}^{-}_{kl}(r)$, that the asymptotic series represents, depends on which route in the complex Z-plane we follow as we change the sign of Z. When the potential is attractive outgoing- and ingoing-wave behavior can be distinguished by the asymptotic series since the oscillations of $\mathscr{R}_{kl}(r)$ are undamped in the region $\sqrt{2Zr/a_0} \ll 2\beta$, and therefore $\mathscr{R}_{kl}(r)$ can be matched to a standing wave in the region of asymptotically large r. Seen another way, since $e^{-2\pi\beta}$ changes from being exponentially small to exponentially large when the sign of Z is changed (with k remaining real and positive) the difference between $\mathscr{I}^{+}_{kl}(r)$ and $\mathscr{I}^{-}_{kl}(r)$ becomes significant. To determine the appropriate branch of the irregular wavefunction for $\mathscr{E}$ real and positive we first put $\arg\mathscr{E} = \pm 2\pi$; thus, before we change the sign of Z, we arrange for $e^{-2\pi\beta}$ to be exponentially large, so that the right side of Eq. (11) represents either $\mathscr{I}^{+}_{kl}(r)$ or $\mathscr{I}^{-}_{kl}(r)$ as discussed in the previous subsection. Subsequently, we let $\arg\mathscr{E}$ revert to 0 and we let the sign of Z change along a route such that β, and hence the branch of the irregular wavefunction, remain unchanged. The upshot is that if we change Z to $e^{-i\pi}Z$ the asymptotic series represents $\mathscr{I}^{+}_{kl}(r)$ while if we change Z to $e^{i\pi}Z$ this series represents $\mathscr{I}^{-}_{kl}(r)$. Thus we arrive at

$$\mathscr{I}^{\pm}_{kl}(r) \sim \pm i\pi \frac{b_l(\beta)}{2C_l(\beta)(2l+1)(kr)^l} \sum_{n=2l+1}^{\infty} (-1)^n a_n(\beta)(2Zr/a_0)^{n/2} H^{\pm}_n(2\sqrt{2Zr/a_0}) \tag{23}$$

where $H^+(z)$ and $H^-(z)$ are Hankel functions of the first and second kind, respectively. Therefore the radial Green function for an *attractive* Coulomb potential has the series representations

$$g_l(r,r',\mathscr{E}\pm i0) \sim \mp i\pi \frac{(2l)!b_l(\beta)}{2Ze^2(2Zr_</a_0)^l(2Zr_>/a_0)^l} \times \left(\sum_{n=2l+1}^{\infty}(-1)^n a_n(\beta)(2Zr_</a_0)^{n/2}J_n(2\sqrt{2Zr_</a_0})\right) \times \left(\sum_{n'=2l+1}^{\infty}(-1)^{n'} a_{n'}(\beta)(2Zr_>/a_0)^{n'/2}H_{n'}^{\mp}(2\sqrt{2Zr_>/a_0})\right) \quad (24)$$

which are applicable when both $r_> \ll r_0$ and $\beta \ll 1$. The right side of Eq. (24) represents the analytic continuation of $g_l(r,r',\mathscr{E}+i0)$ and $g_l(r,r',\mathscr{E}-i0)$ as $\mathscr{E}$ moves into the regions $0 < \arg\mathscr{E} < \pi$ and $0 > \arg\mathscr{E} > -\pi$, respectively. When $|\arg\mathscr{E}| \geq \pi$ the regular wavefunction $\mathscr{R}_{kl}(r)$ is exponentially large if $\sqrt{2Zr_>/a_0} \gg 1$ but merely oscillatory if $\sqrt{2Zr_>/a_0} \ll 1$. Hence the series representations of $g_l(r,r',\mathscr{E}\pm i0)$ are inapplicable along the negative energy axis, where the exact Green function has bound-state poles. Indeed, the bound-state position probablity density peaks well beyond the region $\sqrt{2Zr_>/a_0} \ll 1$ where Eq. (21) is useful.

2.2. *Sturmian Representation*

The Coulomb-Sturmian functions $S_{nl}^k(r)$ are [3]

$$S_{zl}^k(r) = \sqrt{\frac{e^{-i\pi/2k}}{z(z-l)_{2l+1}}}(-2ikr)^{l+1}L_{z-l-1}^{2l+1}(-2ikr)e^{ikr} \quad (25)$$

where $(z-l)_{2l+1}$ a Pochhammer symbol and

$$(2l+1)!L_{z-l-1}^{2l+1}(-2ikr) = (z-l)_{2l+1}\,{}_1F_1(l+1-z,2l+2;-2ikr). \quad (26)$$

The normalization is $\int_0^\infty dr\,[S_{nl}^k(r)]^2 = 1$. The ordinal number (number of nodes) of $S_{nl}^k(r)$ is $n-l-1$. Note that $S_{nl}^k(r)$ is a bound-state eigensolution of the Schrödinger when $\mathscr{E}$ has a fixed real negative value and the potential is $-\lambda_n(Ze^2/r)$ where the overall strength of the potential is the eigenvalue $\lambda_n = -ink/Z$. The radial Green function for an attractive Coulomb potential $(Z>0)$ has the expansion[4]

$$g_l(r,r';\mathscr{E}) = \frac{i}{2\mathscr{E}}\sum_{n=l+1}^{\infty}\left(\frac{nka_0}{Z+inka_0}\right)S_{nl}^k(r)S_{nl}^k(r') \quad (27)$$

with bound-state poles at $ka_0 = iZ/n$, or, equivalently, at $\mathscr{E} = -Z^2e^2/(2n^2a_0)$.

If $\mathscr{E}$ is real and negative the eigenvalue λ_n is real; otherwise it is complex. We can preserve the reality of λ_n for all $\mathscr{E}$ by allowing the ordinal number to become complex. Thus we replace the integer n by an arbitrary complex variable z. To rewrite the series representation of $g_l(\mathscr{E})$ as an integral over z we use a variant of the Sommerfeld-Watson transform. We introduce the weight function[5]

$$w(z) = \left(\frac{1}{2\pi i}\right)\left(\frac{\pi e^{\mp i\pi z}}{\sin\pi z}\right) \quad (28)$$

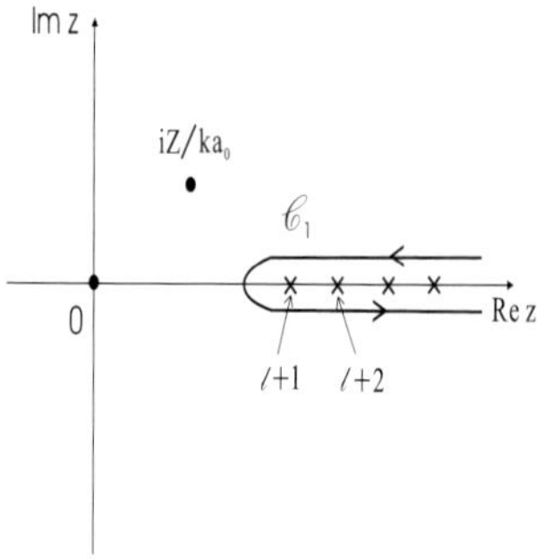

Figure 1. The Sturmian expansion of the Coulomb Green can be transformed from a sum to an integral over a complex variable z. The path of integration, $\mathscr{C}_1$, wraps around the segment $\mathrm{Re}(z) > l$ of the real axis; it encloses simple poles at integer values $l+1, l+2, \ldots$ of z but excludes the pole at $i(Z/ka_0)$.

where the sign of the exponent of $e^{\mp i\pi z}$ is chosen to be minus if $0 \leq |\arg(\mathscr{E})| < \pi$ and plus if $\pi < |\arg(\mathscr{E})| \leq 2\pi$. (The negative real energy axis of the physical sheet, along which $\arg(\mathscr{E}) = \pi$, is excluded.) Since $w(z)$ has poles at integer values of z, each with residue $(1/2\pi i)$, we can replace the sum on the right side of Eq. (27) by an integral over a closed contour $\mathscr{C}_1$ which (see Fig. 1) wraps around the segment $\mathrm{Re}(z) > l$ of the real z-axis and which (assuming $\mathscr{E}$ is not negative) excludes the pole at $i(Z/ka_0)$; we have

$$g_l(r,r';\mathscr{E}) = \frac{i}{2\mathscr{E}}\left(\frac{1}{2\pi i}\right)\int_{\mathscr{C}_1} dz \left(\frac{\pi e^{\mp i\pi z}}{\sin \pi z}\right)\left(\frac{zka_0}{Z+izka_0}\right) S^k_{zl}(r) S^k_{zl}(r'). \tag{29}$$

This integral representation can be analytically continued onto the unphysical sheet by an appropriate distortion of the contour of integration.

The rate $\Gamma/\hbar$ for ionization of a hydrogen atom or hydrogenlike ion by one photon of angular frequency ω and electric field amplitude F is proportional to the imaginary part of the ac quasienergy:

$$\frac{\Gamma}{\hbar} = -2\mathrm{Im}\frac{Z^3F^2e^2}{3a_0^3}\int_0^\infty dr_1 \int_0^\infty dr_2\, e^{-Z(r_1+r_2)/a_0} r_1^2 r_2^2 g_1(r_1,r_2;\mathscr{E}+i0) \tag{30}$$

with $\mathscr{E} = \hbar\omega - \frac{Z^2e^2}{2a_0}$. Using the integral representation, Eq. (29), we can perform the integration over r_1 and r_2. To perform the contour integration we distort the contour so that it follows the right edge of the imaginary z-axis except for an infinitesimal indentation around the pole at iZ/ka_0. The imaginary part of the integral can be evaluated to give

$$\frac{\Gamma}{\hbar} = \frac{4\pi Z^6}{3}\frac{eFa_0}{\hbar}\left(\frac{eFa_0}{\hbar\omega}\right)\left(\frac{e^2}{a_0\hbar\omega}\right)^4\left(\frac{e^{-\frac{4Z}{ka_0}\tan^{-1}\frac{ka_0}{Z}}}{1-e^{-2(Z\pi/ka_0)}}\right) \tag{31}$$

where k is defined by $\hbar\omega = \frac{e^2}{2a_0}(Z^2+k^2a_0^2)$. Due to the Coulomb normalization factor the rate does not vanish as the ionization threshold is approached. As $\mathscr{E}$ approaches zero from above we can use Eq. (24) to give, noting that when x is real $\mathrm{Im}H_n^{\pm}(x) = J_n(x)$,

$$\frac{\Gamma}{\hbar} \sim 2\mathrm{Im}\frac{i\pi F^2Z^2}{3a_0^3}\int_0^\infty dr_1 \int_0^\infty dr_2\, e^{-Z(r_1+r_2)/a_0} r_1^2 r_2^2$$

$$\times (2Zr_</a_0)^{1/2}(2Zr_>/a_0)^{1/2}J_3(2\sqrt{2Zr_</a_0})H_3^+(2\sqrt{2Zr_>/a_0}) \quad (32)$$

$$= \frac{2^4\pi F^2 a_0^3}{3Z^4}\left(\int_0^\infty du\, e^{-u^2}u^6 J_3(\sqrt{8}u)\right)^2 \quad (33)$$

$$= \frac{2^7\pi F^2 a_0^3}{3Z^4\hbar}\exp(-4), \quad \mathscr{E}\sim 0+ \quad (34)$$

in agreement with the limit of Eq. (31) as $\mathscr{E}\to 0+$.

2.3. High-Energy Limit

Let h_l be the radial Hamiltonian. We have $g_l(r,r';\mathscr{E}) = \langle r|g_l(\mathscr{E})|r'\rangle$ where the resolvent $g_l(\mathscr{E})$ is

$$g_l(\mathscr{E}) = \frac{1}{\mathscr{E}-h_l}. \quad (35)$$

We attach superscripts I and II to the resolvent to distinguish the physical and unphysical branches, respectively. At energies on the physical sheet far from the unitarity cut and the bound-state poles of $g_l(\mathscr{E})$ it may be useful to expand the resolvent in an asymptotic power series in $1/\mathscr{E}$:

$$g_l^{(I)}(\mathscr{E}) = \frac{1}{\mathscr{E}}\sum_{n=0}^{\infty}\left(\frac{h_l}{\mathscr{E}}\right)^n. \quad (36)$$

This power series cannot account for the discontinuity between two points that lie on the physical sheet on opposite edges of the cut; the discontinuity is

$$D_l(\mathscr{E}) = -2i\pi\delta(\mathscr{E}-h_l) \quad (37)$$

$$= -2i\pi\rho_l(\mathscr{E})|\mathscr{E}l\rangle\langle\mathscr{E}l| \quad (38)$$

where $|\mathscr{E}l\rangle$ is an eigenket of h_l with continuum-state energy $\mathscr{E}$ and where $\rho_l(\mathscr{E})$ is the density-of-states factor (summed over eigenstates with the same energy eigenvalue). Roughly speaking, we should truncate the power series at a point where the sum dwarfs the discontinuity. Analytic continuation of $g_l(\mathscr{E})$ from the physical sheet to the unphysical sheet can be carried out by means of the relation

$$g_l^{(II)}(\mathscr{E}) = g_l^{(I)}(\mathscr{E}) + D_l(\mathscr{E}) \quad (39)$$

where $D_l(\mathscr{E})$ is the analytic continuation of the discontinuity onto the unphysical sheet.

3. Electron-Electron Repulsion

The interaction between two electrons is $W_{12}\equiv e^2/r_{12}$ where

$$r_{12}^2 = r_1^2 + r_2^2 - 2r_1r_2\cos\theta_{12} \quad (40)$$

with θ_{12} the angle between the electron coordinates $\vec{r}_1$ and $\vec{r}_2$. Let $\underline{M}$ be a matrix representation of r_{12}^2. Since r_{12} is real, symmetric, positive definite, invariant under the parity operation, and scale-invariant, so is both $\underline{M}$ and its square root. Furthermore, the rows and columns of $\underline{M}$ can be labelled so that $\underline{M}$ is tightly banded (e.g. block-diagonal or block-tridiagonal) with respect to the orbital angular momentum quantum numbers of the particles; this yields a substantial reduction in the effort required to diagonalize $\underline{M}$. Once $\underline{M}$ has been reduced to diagonal form it is trivial to find its square root and inverse. However, since $\underline{M}$ is real, symmetric, and positive definite, its square root and inverse can be found without resorting to diagonalization.[8] Instead we can express $\underline{M}$ as the Cholesky decomposition $\underline{L}\,\underline{L}^t$ where $\underline{L}$ is a lower triangular matrix which can be expressed without great effort as the singular-value decomposition $\underline{L} = \underline{X}\,\underline{D}\,\underline{Y}^{\rm t}$ where $\underline{X}$ and $\underline{Y}$ are orthogonal matrices and $\underline{D}$ is diagonal. Since $\underline{L}\ underline L^t = (\underline{X}\,\underline{D}\,\underline{Y}^{\rm t})(\underline{Y}\,\underline{D}\,\underline{X}^{\rm t}) = (\underline{X}\,\underline{D}\,\underline{X}^{\rm t})^2$ we have $\underline{M}^{-1/2} = \underline{X}\,\underline{D}^{-1}\,\underline{X}^{\rm t}$.

Table 1. Estimates of the 1^1S-2^1P oscillator strength f of He in atomic units, calculated within the length gauge for different sizes of a Sturmian basis whose wavenumber (inverse length scale) is κ. The characteristic range of the basis is n/κ, and the maximum number of nodes of a basis function is $\tilde{n} \equiv n-l-1$, where n and l are the maximum values of the ordinal and angular momentum quantum numbers employed in the basis. A more accurate value [7] of the 1^1S-2^1P oscillator strength is 0.27616.

$\tilde{n}$	l	$f(\kappa=1.0)$	$f(\kappa=1.5)$	l	$f(\kappa=1.0)$	$f(\kappa=1.5)$	l	$f(\kappa=1.0)$	$f(\kappa=1.5)$
5	5	0.24141	0.30832	7	0.24138	0.30812	9	0.24138	0.30811
10	5	0.26662	0.27249	7	0.26646	0.27194	9	0.26645	0.27187
15	5	0.27082	0.27401	7	0.27044	0.27319	9	0.27041	0.27304
20	5	0.27283	0.27515	7	0.27223	0.27419	9	0.27215	0.27395
25	5	0.27403	0.27581	7	0.27325	0.27481	9	0.27312	0.27450
30	5	0.27483	0.27622	7	0.27393	0.27524	9	0.27374	0.27487

Table 2. Same as table 1 but for the 2^1P-3^1D oscillator strength; a more accurate value [7] is 0.71016.

$\tilde{n}$	l	$f(\kappa=1.0)$	$f(\kappa=1.5)$	l	$f(\kappa=1.0)$	$f(\kappa=1.5)$	l	$f(\kappa=1.0)$	$f(\kappa=1.5)$
5	5	0.81497	0.88433	7	0.81500	0.88436	9	0.81500	0.88436
10	5	0.71069	0.81303	7	0.71084	0.81351	9	0.71084	0.81357
15	5	0.70630	0.72431	7	0.70664	0.72506	9	0.70667	0.72520
20	5	0.70697	0.70885	7	0.70750	0.70974	9	0.70757	0.70996
25	5	0.70731	0.70771	7	0.70801	0.70870	9	0.70813	0.70898
30	5	0.70749	0.70769	7	0.70831	0.70874	9	0.70848	0.70907

While matrix elements of $1/r_{12}$ calculated in this way are not plagued by roundoff error they do suffer from truncation error due to the truncation of the basis. Since r_{12}^2 weights more heavily the region of large r_{12}, the inter-particle interaction is not very accurately described in the region where the ground- and lowly-excited bound-state wavefunctions peak in magnitude. In tables 1 and 2 we show estimates of the 1^1S-2^1P and 2^1P-3^1D oscillator strengths of He for different basis sizes and two different length scales. The convergence with increasing ordinal number is slow, but is more rapid for the shorter length scale.

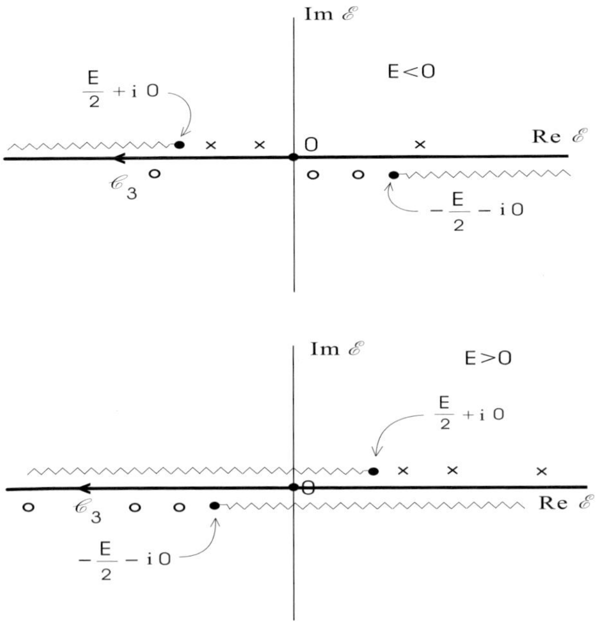

Figure 2. The path of integration of the convolution integral is the whole real $\mathscr{E}$ axis, from ∞ to $-\infty$. The bound-state poles which are associated to the thresholds at $\frac{1}{2}E$ and $-\frac{1}{2}E$ are shown schematically as crosses and open circles, respectively. The zig-zag lines are the *unitarity* branch cuts. If $\mathscr{E}_{\rm gd}$ is the ground-state energy of the one-electron system, the two-electron system is unstable if $E > \mathscr{E}_{\rm gd}$, i.e. if the ground-state pole of one of the one-particle Green functions sits on the edge of the unitarity cut of the other one-particle Green function. Bound- and quasibound-state poles of the full two-particle resolvent $G(E)$ develop from degeneracies, i.e. from discrete points on the real $\mathscr{E}$-axis where bound-state poles of different one-particle Green functions coincide.

4. Convolution of Two One-Particle Green Functions

Let E be the total energy of two electrons, 1 and 2 say, moving under the influence of both their mutual interaction W_{12} and a fixed center of force; E is real, but may be positive or negative. If $G_0(E)$ is the resolvent for the motion in the absence of W_{12}, the full two-particle resolvent is

$$G(E) = [W_{12} - W_{12}G_0(E)W_{12}]^{-1}W_{12}G_0(E). \tag{41}$$

Since $[W_{12} - W_{12}G_0(E)W_{12}]$ is symmetric and positive definite on the negative real energy axis below the bound-state poles of $G_0(E)$ (note W_{12} is positive definite) $[W_{12} - W_{12}G_0(E)W_{12}]$ has a Cholesky decomposition,[8] which can be analytically continued to other regions of the complex E-plane, in particular to either edge of the unitarity cut. Thereby the inverse of $[W_{12} - W_{12}G_0(E)W_{12}]$ can be calculated efficiently and with high numerical stability.

Let $g(1;\mathscr{E})$ and $g(2;\mathscr{E})$ be the one-particle resolvents for particles 1 and 2 to move on their own in the presence of the center of force (the nucleus). Introducing $\mathscr{E}_0 \equiv \frac{1}{2}E$ the physical branch of $G_0(E)$ is the standard convolution integral,[1] as yet unsymmetrized with respect to particle exchange,

$$G_0^{(I)}(E+i0) = \frac{1}{2\pi i}\int_{\mathscr{C}_3} d\mathscr{E}\, g^{(I)}(1;\mathscr{E}_0+i0-\mathscr{E})g^{(I)}(2;\mathscr{E}_0+i0+\mathscr{E}) \tag{42}$$

where $\mathscr{C}_3$ runs along real axis from ∞ to $-\infty$ and where the superscript I indicates that the branch is the physical one. The contour $\mathscr{C}_3$ runs just beneath the cut and the bound-state poles of $g^{(I)}(1;\mathscr{E}_0+i0-\mathscr{E})$, and just above the cut and the bound-state poles of $g^{(I)}(2;\mathscr{E}_0+i0+\mathscr{E})$; see Fig. 2. The branch points and bound-state poles can be avoided by rotating the contour about the origin through an in the range $[-\pi/2,\pi/2]$. If E is negative the rotation is counterclockwise; the contour remains on the physical sheet but sweeps past those poles further than $|\mathscr{E}_0|$ from threshold. If E is positive the rotation is clockwise; the contour sweeps over the unitarity cuts and moves onto the unphysical sheet. When we symmetrize $G_0(E)$ with respect to particle exchange the symmetrized integrand is even in $\mathscr{E}$, and so we can reduce the range of integration to $[0,\infty]$. If E is positive we have (taking into account that the contour sweeps over the unitarity cuts)

$$\begin{aligned} G_0^{(I)}(E>0+i0) &= -\frac{1}{2\pi i}\int_0^{\infty} d\mathscr{E}\,\Big[g^{(I)}(1;\mathscr{E}_0-\mathscr{E})g^{(II)}(2;\mathscr{E}_0+\mathscr{E}) \\ &\quad + g^{(II)}(1;\mathscr{E}_0+\mathscr{E})g^{(I)}(2;\mathscr{E}_0-\mathscr{E})\Big], \quad \arg\mathscr{E}<0 \end{aligned} \tag{43}$$

where the superscript II indicates that the branch is the unphysical one. If E is negative we have (taking into account that the contour sweeps over the poles)

$$\begin{aligned} G_0^{(I)}(E<0) &= -\frac{1}{2\pi i}\int_{-\infty}^{0} d\mathscr{E}\,\Big[g^{(I)}(1;\mathscr{E}_0-\mathscr{E})g^{(I)}(2;\mathscr{E}_0+\mathscr{E}) \\ &\quad + g^{(I)}(1;\mathscr{E}_0+\mathscr{E})g^{(I)}(2;\mathscr{E}_0-\mathscr{E})\Big] + R(E), \quad \arg\mathscr{E}>0, \end{aligned} \tag{44}$$

where $R(E)$ is the sum of the residues of those bound-state poles crossed by the contour, i.e.

$$R(E) = \sum_{\mathscr{E}_{\rm bd}<\mathscr{E}_0} g^{(I)}(1;E-\mathscr{E}_{\rm bd})|2;\mathscr{E}_{\rm bd}\rangle\,\langle 2;\mathscr{E}_{\rm bd}| \;+\; |1;\mathscr{E}_{\rm bd}\rangle\,\langle 1;\mathscr{E}_{\rm bd}|g^{(I)}(2;E-\mathscr{E}_{\rm bd}) \tag{45}$$

where $|j;\mathscr{E}_{\rm bd}\rangle$ is an eigenket of the Hamiltonian $h(j)$ for the j-th particle with bound-state en ergy $\mathscr{E}_{\rm bd}$. When $E<0$ degeneracies can occur at those points for which $E=\mathscr{E}_{\rm bd}+\mathscr{E}'_{\rm bd}$, where $\mathscr{E}_{\rm bd}$ and $\mathscr{E}'_{\rm bd}$ are bound-state energies of particles 1 and 2, respectively, with $\mathscr{E}_{\rm bd}\le\mathscr{E}_0$ and $\mathscr{E}'_{\rm bd}\ge\mathscr{E}_0$. These degeneracies give rise to poles of $R(E)$, which, through the coupling W_{12}, lead to bound-state (if $\mathscr{E}_{\rm gd}>\mathscr{E}_0$) and quasibound-state (if $\mathscr{E}_{\rm gd}<\mathscr{E}_0$) poles of the full two-particle resolvent $G(E)$ (where $\mathscr{E}_{\rm gd}$ is the ground-state energy of the one-particle system). The poles of $R(E)$ can be removed by replacing the Hamiltonian $h(j)$ for the j-th particle ($j=1,2$) by $P(j)h(j)P(j)$ where $P(j)$ is the projection operator whose complement $Q(j)=1-P(j)$ projects onto those eiegenstates of $h(j)$ with eigenvalues in the interval $[E,\frac{1}{2}E]$. Accordingly, the "perturbation" $Q(j)h(j)Q(j)$ must be added to W_{12}. The rotation angle of the contour of integration must be sufficiently large that it sweeps past the (apriori undetermined) quasibound-state poles of $G(E)$.

The rotated straight-line integration contour penetrates deep into the region of (complex) unphysical energies where numerical stability deteriorates. It is more helpful to use a contour which remains close to the real energy axis except near the singularities. This can be accomplished by invoking the conformal transformation

$$u = \frac{\mathscr{E}_0-\mathscr{E}}{\mathscr{E}_0+\mathscr{E}} \tag{46}$$

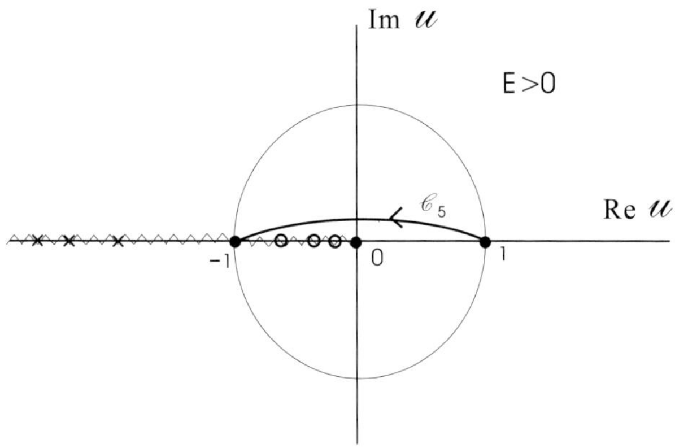

Figure 3. Half of the energy plane is mapped onto the disc $|u| \leq 1$. The integration contour $\mathscr{C}_5$ runs from $u = 1$ to $u = -1$, and avoids the high-lying bound-state poles which accumulate below $u = 0$. This illustration is for $E > 0$; if $E < 0$ the integration contour runs from $u = -1$ to $u = 1$ in the lower half of the disc.

which maps half of the energy plane onto the disc $|u| \leq 1$. We have

$$\mathscr{E}_0 - \mathscr{E} = \left(\frac{u}{1+u}\right)E, \quad \mathscr{E}_0 + \mathscr{E} = \left(\frac{1}{1+u}\right)E, \tag{47}$$

and $d\mathscr{E} = -Edu/(1+u)^2$. The contour $\mathscr{C}_5$ in the u-plane (see Fig. 3) commences (terminates) at $u = 1$ and terminates (commences) at $u = -1$ if $E > 0$ (< 0); it remains close to the real u-axis except in the region of the origin where it circumvents the branch point at $u = 0$ and also the high-lying bound-state poles which accumulate to the left of this branch point.

5. Numerical Application

We have tested the method outlined above by applying it to the problem of one-photon absorption by H^-. We treated the nucleus as an infinitely massive and stationary center of force. Our basis was composed of symmetrized products of real one-particle Sturmian functions. The calculations were performed on a single-processor (Pentium 4, 1.7GHz) desktop with 1.25Gb of internal memory.

Our calculations were done in the velocity gauge. We took the maximum ordinal number of the Sturmian functions to be 30 (25), and the maximum angular momentum quantum number to be 5 (7), at photon energies below (above) 18 eV. In Fig. 4 we show results for the ratio of the cross sections for double- and single-electron escape over a range of photon energies above the threshold for double-escape. We show both our results and those of Kheifets and Bray,[9]. The results of Kheifets and Bray agree well with the earlier results (not shown) of Meyer, Greene, and Esry[10] and Edah, Foumouo, and Piraux[11]. Our results are not sufficient in accuracy to confirm the Wannier law $\sigma = \sigma_0(E - E_0)^{1.127}$ for the double escape cross section in the region just above the energy threshold E_0; but if we accept this law we find that σ_0 is roughly 100 kb when the energy E is expressed in units of eV, a value which is of the same order of magnitude as the value 95 kb found by Kheifets and Bray [9].

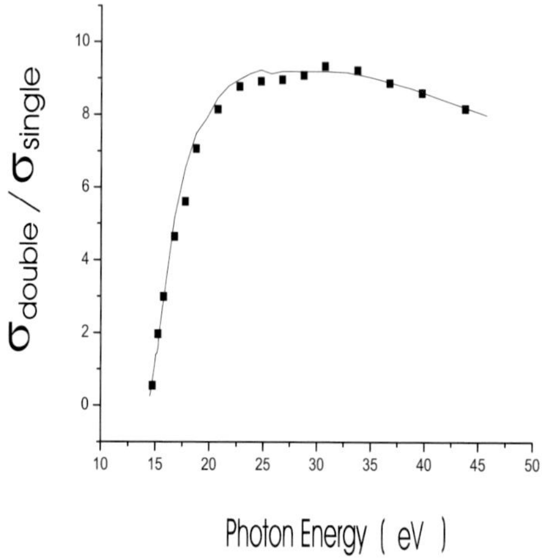

Figure 4. The ratio of cross sections for double- and single-electron escape from H^- by one-photon absorption. The solid curve corresponds to the results of Kheifets and Bray,[9]. The solid squares are the present results.

Acknowledgments

I gratefully acknowledge support from the NSF, under Grant No. PHY-0245290.

References

[1] Papp Z, Darai J, Hlousek Z T, Konya B, and Yakolev S L 2002 *Phys. Rev.* A**65** 032725

[2] Abramowitz M and Stegun I E 1965 *Handbook of Mathematical Functions* (U.S. Dept. of Commerce, Washington D.C.)

[3] Rotenberg M 1970 *Adv. At. Mol. Phys.* **6** 233

[4] Hostler L C 1970 *J. Math. Phys.* **11** 2966

[5] Shakeshaft R 2004 *Phys. Rev.* A**70** 042704

[6] Shakeshaft R 1999 *Phys. Rev.* A**60** 1280; *ibid* 4198

[7] Kono A and Hattori S 1985 *Phys. Rev.* A**31** 1199

[8] Golub G H and Van Loan C F 1996 *Matrix Computations*, 3rd ed. (John Hopkins, Baltimore)

[9] Kheifets A and Bray I 1998 *Phys. Rev.* A **58**, 4501

[10] Meyer K W, Greene C H, and Esry B D 1997 *Phys. Rev. Lett.* **78** 3662

[11] Edah G, Foumouo E, and Piraux B 2003 (unpublished)

Inst. Phys. Conf. Ser. No. 183
Paper presented at Int. Conf. Electron and Photon Impact Ionization and Related Topics, Louvain-la-Neuve, 2004

Interaction of strong X-ray lasers with small argon clusters

Ulf Saalmann

Max-Planck-Institut für Physik komplexer Systeme
Nöthnitzer Str. 38, 01187 Dresden, Germany

Abstract. In theoretical studies of intense X-ray laser impact on small argon clusters we observed a suppression of ionization for clusters compared to that of single atoms. This is in contrast to experimental and theoretical investigations of cluster ionization in intense optical laser pulses. The underlying mechanisms of laser-cluster interaction at X-ray wavelengths are revealed by studying the time-dependent ionization and shell-specific contributions to the total ionization yield. Additional insight is obtained by varying pulse length and frequency of the laser.

1. Introduction

There has been a renewed interest in laser-matter interaction since strong femtosecond laser sources have become available. In contrast to the weakly perturbing laser light in spectroscopic investigations, the *high intense* fields from these ultrashort pulses may induce novel, non-linear processes in atomic systems [1]. In particular, atomic clusters have been studied extensively [2–7] to understand more generally the energy transfer from intense laser light into complex many-particle targets. So far these studies are almost exclusively restricted to the visible or infrared wavelength regime with only one series of experiments which reports Coulomb explosion of rare-gas clusters in intense VUV laser light [7]. The new X-ray free electron laser (XFEL) sources, e. g. those under construction at DESY in Hamburg [8] and at the LCLS in Stanford [9], will change this situation. They can deliver intense laser pulses at *high frequencies* (from VUV to hard X-ray). This will not only open a new regime of strong-field atomic physics but also allow for novel applications like, e. g., imaging methods to get structural and dynamical information with atomic resolution on a femtosecond time scale [8].

To elucidate the impact of intense X-ray laser light on atomic clusters we have performed theoretical investigations for small argon clusters using a microscopic description based on a mixed quantum/classical treatment of the electron and ion dynamics. The laser parameter are as follows: frequency $\hbar\omega = 350\ldots500\,\mathrm{eV}$, i. e. wavelength $\lambda \approx 3.5\ldots2.5\,\mathrm{nm}$, intensity $I \approx 10^{14}\ldots10^{18}\,\mathrm{W/cm^2}$, pulse length $T = 100\ldots200\,\mathrm{fs}$. Thus we are in the regime of interest for upcoming XFEL machines [8, 9].

2. Theoretical

2.1. Important scales and basic mechanisms

In the short-wavelength ($\lambda \sim 1$ nm) regime the interaction of laser light with atoms is notably different from long-wavelength ($\lambda \sim 1\,\mu$m) pulses which is already evident from the motion of free electrons driven by a laser field with intensity I. This oscillatory motion can be characterized by the ponderomotive energy $E_{\mathrm{pond}} \sim I\lambda^2$, the average kinetic energy, and the quiver amplitude $x_{\mathrm{quiv}} \sim \sqrt{I}\lambda^2$, the spatial excursion in the laser field. Note that both quantities scale quadratically with the wavelength λ. Table 1 compares values for an optical laser to those for an X-ray laser studied below.

wavelength	intensity $I=$	10^{14} W/cm^2	10^{16} W/cm^2	10^{18} W/cm^2
$\lambda = 780$ nm	x_{quiv}	8.28Å	82.8 Å	828 Å
	E_{pond}	5.67 eV	567 eV	56.7 keV
$\lambda = 3.5$ nm	x_{quiv}	0.0002 Å	0.002 Å	0.02 Å
	E_{pond}	0.1 meV	0.01 eV	1.1 eV

Table 1. Quiver amplitudes x_{quiv} and ponderomotive energies E_{pond} for long and short laser wavelengths λ at different intensities I.

The small ponderomotive energies and the small quiver amplitudes at short wavelengths have two important consequences:

(1) Despite the high intensities the laser-atom interaction is of non-relativistic and perturbative nature. The first is obvious as $E_{\mathrm{pond}} \ll E_0$, with $E_0 \approx 511$ keV the rest mass energy of the electron. The latter is also clear from the so-called Keldysh parameter γ which gives the ratio between the tunneling time and the laser period [10] and can be rewritten in terms of the binding energy E_{bind} and the ponderomotive energy E_{pond} as $\gamma = \sqrt{E_{\mathrm{bind}}/2E_{\mathrm{pond}}}$. For typical binding energies of a few 10 or 100 eV we find $\gamma \gg 1$. Obviously the laser period is far too short for tunnel-ionization. Rather, ionization is due to single-photon absorption, a first-order process in a perturbative treatment.

(2) Forces on free or quasi-free[1] electrons due to the laser can be completely neglected. This indicates insignificant excitation of quasi-free electrons by inverse bremsstrahlung, a crucial heating mechanism of clusters in optical lasers [11].

From the above should be clear that ionization proceeds fundamentally different under the laser conditions mentioned in Sect. 1 from that under optical laser impact. Ionization starts from the inside because photo-ionization cross sections σ at X-ray wavelengths are considerably higher for the inner shells than for the valence shells [12]. In first-order perturbation theory cross sections scale as $\sigma \propto E_{\mathrm{bind}}{}^{7/2}$ for $\hbar\omega > E_{\mathrm{bind}}$. Typically, the inverse rates are with $1\ldots10$ fs much smaller than the pulse length of about 100 fs.

[1] Electrons which are excited from their initial bound state but are still bound by the cluster potential as a whole. They can move freely within the cluster volume.

Hence, multiple single-photon ionization is possible, in particular because the inner-shell holes created by photo-ionization are refilled by Auger-like processes. The Auger decay is only weakly dependent on the atomic charge state [13] and occurs fast, typical times are $0.2\ldots5\,\mathrm{fs}$ [14]. Due to this almost instantaneous refilling of the inner shells they can be ionized many times during the pulse and thus the atoms can be efficiently "pumped dry". This occurs "inside-out" and is the exact opposite to the ionization mechanism in the visible wavelengths regime where the most weakly bound electrons are removed first. It should be mentioned that this ionization cascade may stop for highly charged ions where the increasing binding energy of the remaining electrons may prevent both photo- and auto-ionization for energetical reasons.

Finally, local ionic charges in the cluster, generated in the course of the ionization process, may be screened by the weakly bound valence electrons of neighbouring cluster atoms. Therefore, these electrons may tunnel through the barriers between the mother atom/ion and neighbouring ions and thus delocalize. Such electrons will predominantly move *inside* the cluster since there is no field (such as the quasi-static electric field of low-frequency lasers) which would drive them away from the charged cluster into the continuum.

2.2. Model

In order to gain insight into the laser induced dynamics we use a mixed quantum/classical approach similar to those successfully applied in studying rare-gas clusters in strong pulses of optical light [11, 15–17]. The key point is the subdivision of bound and (quasi-)free electrons. The dynamics of bound electrons with typically oscillation periods of a few attoseconds are not treated explicitly. Rather we use a statistical approach to describe it by means of the occupation number of bound levels which may change after each time step. The probability for a particular transition within a time step is calculated as the product of the corresponding rate Γ and the time step Δt. This probability $p = \Gamma \cdot \Delta t$ is compared to a random number ξ distributed uniformly in the interval [0,1], a transition takes place if $p > \xi$. The rates Γ may crucially depend on the laser (intensity and frequency) and the current state of ion, cf. Sect. 2.1. The photo-ionization can be calculated by first-order perturbation theory [18] or using parameterized cross section [19]. For the intra-atomic decay cascades we use direct Hartree-Fock calculations which provide branching ratios and decay rates for Auger, Coster-Kroning, and radiative transitions [14]. To incorporate the above mentioned screening we allow for intra-cluster tunneling. The rates are calculated by means of the tunneling integral in direction of the electric field at the atom under consideration [17].

An inner ionization event gives birth to a quasi-free electron which is subsequently propagated classically along with the ions and other quasi-free electrons with all mutual Coulomb forces included. Note that that probabilities of these electrons for photoabsorption and auto-ionizing collisions with bound electrons are considerably smaller than for the bound electrons and thus neglected [20]. Radiative recapture processes are unlikely because of their low recombination rates [18]. We emphasize here again that a quasi-free electron is not necessarily ionized with respect to the cluster as a whole, quasi-free means only inner-ionized with respect to the mother atom or ion.

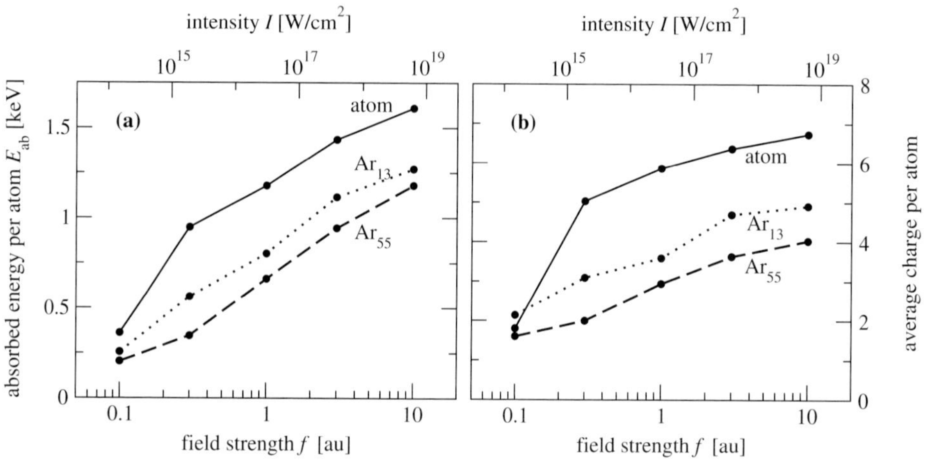

Figure 1. Absorbed energy E_{ab} per atom **(a)** and final average charge per atom **(b)** for two cluster sizes Ar_{13} and Ar_{55} produced by an XFEL pulse ($\hbar\omega = 350\,\mathrm{eV}$, $T = 100\,\mathrm{fs}$) as a function of the field strength f compared to atomic argon as target. Each point is the average of 10 simulations.

3. Results for cluster ionization

3.1. Clusters versus atoms

Here we present results for argon clusters in strong XFEL pulses. To start with we compare the energy absorption and ionization of small clusters with those of atoms [20]. Figure 1 shows the absorbed energy and the final charge divided by the number of atoms in the clusters for a photon energy of $\hbar\omega = 350\,\mathrm{eV}$. Whereas the charge distribution of the fragments, and thus the average charge, is directly measurable in the experiment, the absorbed energy can only be deduced from the calculation. It is shown here since it gives direct information on the number of photons absorbed during the pulse while all the other processes (intra-atomic decay, intra-cluster screening) are not influenced by the laser.

At low field strengths $f = 0.1$ au, only about one single photo-ionization event per atom occurs independently of the cluster size, the absorbed energy per atom is $E_{ab} \sim 350\,\mathrm{eV}$. This energy E_{ab} increases with the laser field strength to more than 1 keV (i. e. about 3 to 4 photoionized electrons per atom) for $f = 10$ au. However, values for the clusters are always below those of the atom. This reduction is stronger for the larger cluster. It indicates a considerable influence of the cluster environment on the photo-ionization process. Firstly, this was unexpected because photo-ionization is predominantly an inner-shell process and possible effects on the photo-ionization rates of these strongly localized electrons from neighbouring ions are unlikely. See, however, the discussion in Sect. 3.2 below.

The final charge of the cluster ions, cf. Fig. 1(b), accounts besides the photo-ionization also for intra-atomic decays and "post-collisional effects" by propagation of the inner-ionized electrons. Apart from the quite steep rise of the final charge in the atomic case for low fields $f < 0.3$ au, the overall increase with the field strength f is weaker in particular for the cluster. Thus — in striking contrast to optical laser-cluster interaction — clusters are less effectively ionized at *high* fields than atoms. Again the reduction is more pronounced for the larger cluster.

3.2. Reduced ionization in clusters

The reason for the reduced ionization in the cluster compared to the atom is twofold:

(i) High fields produce a much larger space charge in a cluster. Such high space charges suppress ionization, because the absorption of one single photon transfers only a fixed amount of energy to the electron. The rapidly oscillating field is unable to drive electrons against the positive space charge out of the cluster which is also evident from the small quiver amplitudes x_{quiv} listed in Table 1.

(ii) The intra-cluster screening leads to a delocalization of the valence electrons which results in two effects: Firstly, the photo-ionization cross sections become very small since the electrons itselves are far away from the nucleus. Secondly, also the Auger decay rates are reduced because the overlap with the core holes becomes smaller.

Whereas (i) follows directly from the propagation of the electrons and ions, (ii) emerges due to the neglection of photo- and auto-ionization of quasi-free electrons. This is justified by the scaling of ionization rates with binding energies which are considerably smaller for quasi-free electrons than those of electrons in bound atomic states [20].

The effect of intra-cluster screening can be quantified by artificially excluding tunneling of electrons to neighbouring ions. Figure 2 shows the result of this (right column) in comparison to a full calculation (left column). Clearly, this artificial constriction of electrons to

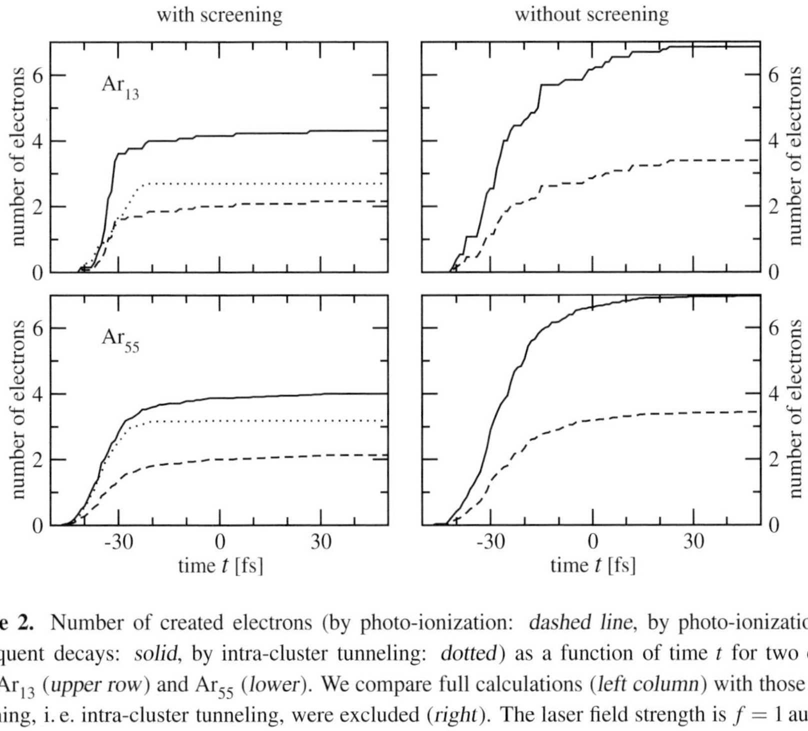

Figure 2. Number of created electrons (by photo-ionization: *dashed line*, by photo-ionization and subsequent decays: *solid*, by intra-cluster tunneling: *dotted*) as a function of time t for two cluster sizes Ar_{13} (*upper row*) and Ar_{55} (*lower*). We compare full calculations (*left column*) with those where screening, i. e. intra-cluster tunneling, were excluded (*right*). The laser field strength is $f = 1$ au, other laser parameters as in Fig. 1.

bound states which corresponds to a localization of the electrons around the ionic cores increases their ionization rates. Note that this applies to primary photo-ionization and secondary auto-ionization processes as well.

For the full calculation (left column of Fig. 2) the final number electrons is reached relatively early in the pulse. To understand this one may look at the contributions from particular shells of the atoms. Figure 3 shows the contributing and competing processes resolved for the different shells of argon. The photo-ionization of the L-shell is suppressed when the binding energy falls below the single-photon threshold. This occurs earlier in the case of clusters due to the valence delocalization which is connected with a reduced electron-electron repulsion within the atom. This faster increase of the L-shell binding energies in clusters explains both the weaker ionization compared to atoms (lower left part of Fig. 3) and the earlier saturation in the pulse (left column of Fig. 2). The latter follows from the fact that excluding tunneling makes the atoms in the cluster behave like separated atoms (apart from space charge effects).

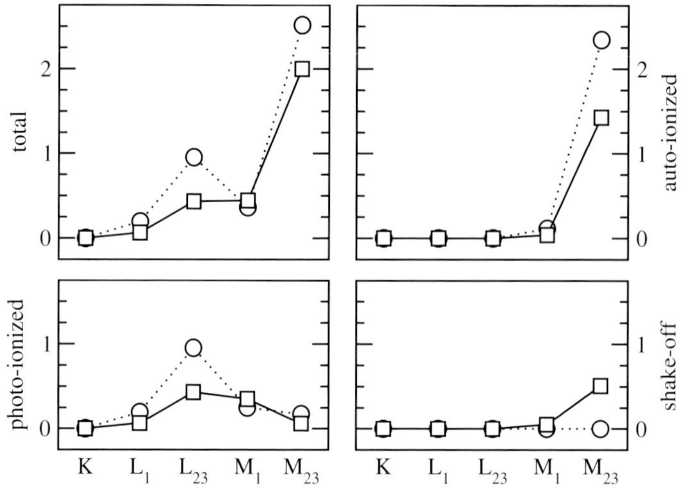

Figure 3. Number of electrons per atom for Ar_{55} due to photo-ionization, auto-ionization, shake-off process and their total sum with respect to the electronic shell they originated from. As in Fig. 2, full calculations including screening (*square symbols*) are compared to calculations excluding screening (*circles*).

One should emphasize that Fig. 3 counts only those electrons which are finally free, i. e. have a positive energy, at the end of the pulse. This explains that the dominating sub-shell for photo-ionization is the L_{23} with a smaller binding energy than L_1. In an analogous manner only electrons from the most weakly bound sub-shell M_{23} acquire in a two-electron Auger process enough energy to leave the cluster. Furthermore, Fig. 3 shows that the delocalization effect is of comparable importance for both photo-ionization and auto-ionization.

3.3. *Pulse-length and frequency dependence*

Experimental [5, 6] and theoretical [11] investigations of clusters in *optical* laser fields have revealed a crucial dependence of the cluster ionization on the pulse length. The reason is a resonance effect, where the eigenfrequency of the cluster potential (i. e. the attractive potential

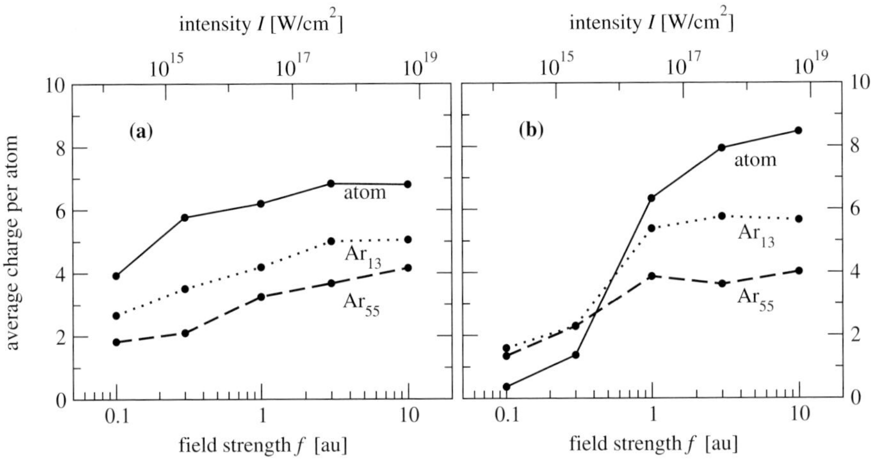

Figure 4. Average charge per atom for two cluster sizes Ar_{13} and Ar_{55} produced by an XFEL pulse as a function of the field strength f for two different laser pulses: **(a)** $\hbar\omega = 350\,\text{eV}$, $T = 200\,\text{fs}$ and **(b)** $\hbar\omega = 500\,\text{eV}$, $T = 100\,\text{fs}$.

of the ions) coincides with the laser frequency, which leads to effective heating and ionization [11]. This frequency matching becomes possible due to the fast expansion of the charged cluster on a femtosecond time scale, i. e. typically during the pulse. To study the influence of the pulse length in X-ray lasers we have repeated the calculations shown Fig. 1(b) for a pulse twice as long, $T = 200\,\text{fs}$. As can be seen in Fig. 4(a) the behaviour is quite similar, i. e. independent on the pulse length, in particular towards higher field strengths $f \gtrsim 1$ au. For low fields $f \sim 0.3$ au, the ionization of the atoms increases with the longer interaction time. This effect is considerably weaker for the cluster. A resonant heating cannot be observed since the laser frequency ($> 100\,\text{eV}$) is far above the eigenfrequency of the cluster potential ($\sim 1\,\text{eV}$).

Additionally, one may study the influence of the laser frequency which will be freely adjustable for XFEL machines [8]. If one uses higher-frequency photons the picture reverses for the low fields as can be seen in Fig. 4(b) for $\hbar\omega = 500\,\text{eV}$. For these parameters the delocalization of valence electrons, connected with an decreasing of the inner-shell binding energies, favours photo-ionization since the cross sections becomes larger. The multiple ionization for higher fields, however, shows the same behaviour observed for $\hbar\omega = 350\,\text{eV}$. The steeper rise for the atom is due to strong non-linear dependence on the binding energy, cf. Sect. 2.1. Therefore the transition from only weak ionization to the saturation regime is much faster than for clusters where the intra-cluster effects leads to a weakening of the change of binding energy (or localization) of bound electrons.

4. Summary

Our theoretical investigations of small argon clusters subjected to strong X-ray laser pulses showed that the average charge per atom, which characterizes the amount of energy absorbed by the cluster, depends strongly on the cluster size. The lower charges of the ions from the larger clusters can be explained by two effects: the higher space charge connected with a small quiver amplitude of the laser and the delocalization of valence electrons resulting in lower ionization rates. This picture changes only slightly for longer pulses. However, for particular combinations of laser frequency ω and intensity I one may reverse the situation, e. g. for $\hbar\omega = 500\,\mathrm{eV}$ the fragment charges of the cluster are higher than those of single atoms for intensities $I < 10^{16}\,\mathrm{W/cm^2}$.

Many helpful discussions with Jan-Michael Rost are gratefully acknowledged

References

[1] *Super-Intense Laser-Atom Physics*, edited by B. Piraux and K. Rzazewski (Kluwer Academic, Dordrecht, 2001).

[2] T. Ditmire *et al.*, Nature **386**, 54 (1997).

[3] Y. L. Shao *et al.*, Phys. Rev. Lett. **77**, 3343 (1996).

[4] A. McPherson *et al.*, Nature **370**, 631 (1994).

[5] J. Zweiback, T. Ditmire, and M. D. Perry, Phys. Rev. A **59**, R 3166 (1999).

[6] L. Köller *et al.*, Phys. Rev. Lett. **82**, 3783 (1999).

[7] H. Wabnitz, L. Bittner, A. R. B. de Castro, R. Döhrmann, P. Gürtler, T. Laarmann, W. Laasch, J. Schulz, A. Swiderski, K. von Haeften, T. Möller, B. Faatz, A. Fateev, J. Feldhaus, Ch. Gerth, U. Hahn, E. Saldin, E. Schneidmiller, K. Sytchev, K. Tiedtke, R. Treusch, and M. Yurkov, Nature (London) **420**, 482 (2002).
T. Laarmann, A. R. B. de Castro, P. Gürtler, W. Laasch, J. Schulz, H. Wabnitz, and T. Möller, Phys. Rev. Lett. **92**, 143401 (2004).

[8] The X-ray free electron laser, Vol. V in *TESLA Technical Design Report*, edited by G. Materlik and T. Tschentscher (DESY, Hamburg, 2001).

[9] J. Arthur, Rev. Sc i. Instrum. **73**, 1393 (2002).

[10] L. V. Keldysh, Sov. Phys. JEPT **20**, 1307 (1965).

[11] U. Saalmann and J.-M. Rost, Phys. Rev. Lett. **91**, 223401 (2003).

[12] M. Y. Amusia, *Atomic photoeffect* (Plenum Press, New York, 1990).

[13] For the idealized case of hydrogenic wave-functions it can be shown that the matrix element for Auger decay according to Fermi's golden rule is completely independent of the nuclear charge.

[14] A. G. Kochur, V. L. Sukhorukov, A. J. Dudenko, and P. V. Demekhin, J. Phys. B **28**, 387 (1995).

[15] C. Rose-Petruck, K. J. Schafer, K. R. Wilson, and C. P. J. Barty, Phys. Rev. A **55**, 1182 (1997).

[16] K. Ishikawa and T. Blenski, Phys. Rev. A **62**, 063204 (2000).

[17] Ch. Siedschlag and J. M. Rost, Phys. Rev. Lett. **89**, 173401 (2002); Phys. Rev. A **67**, 013404 (2003).

[18] *Quantum mechanics of one- and two-electron atoms* by H. A. Bethe and E. E. Salpeter (Plenum Press, New York, 1957).

[19] D. A. Verner, D. G. Yakovlev, I. M. Band, and M. B. Trzhaskovskaya, At. Data Nucl. Data Tabl. **55**, 233 (1993).

[20] U. Saalmann and J.-M. Rost, Phys. Rev. Lett. **89**, 143401 (2002).

Inst. Phys. Conf. Ser. No. 183
Paper presented at Int. Conf. Electron and Photon Impact Ionization and Related Topics, Louvain-la-Neuve, 2004

Electron impact detachment of small negative clusters

D. Zajfman† ‡, O. Heber†, A. Diner†, P. Witte† ‡, D. Strasser†, Y. Toker†, M. L. Rappaport§, I. Ben-Itzhak† ||[1], O. Guliamov¶, L. Kronik¶, D. Schwalm‡, A. Wolf‡

† Department of Particle Physics, Weizmann Institute of Science, Rehovot, 76100, Israel

‡ Max-Planck-Institut für Kernphysik, D-69117 Heidelberg, Germany

§ Physics Services Unit, Weizmann Institute of Science, Rehovot, 76100, Israel

||J.R. Macdonald Laboratory, Physics Department, Kansas State University, Manhattan, KS 66506 USA

¶Department of Materials and Interfaces, Weizmann Institute of Science, Rehovot, 76100 Israel

Abstract. Results for electron impact detachment of internally cold C_n^- and Al_n^- clusters are presented. The experiments were performed using an electrostatic ion beam trap fitted with an internal electron target. The measured electron impact detachment cross sections for the C_n^- (n=1-9) clusters exhibit an even-odd oscillation, which is in agreement with the familiar even-odd variation in the binding energy, namely higher cross section for weaker binding. However, the cross sections increase on average with the cluster size, in spite of the average increase in the binding energy. We suggest, based on the data and on preliminary calculations, that the size dependent polarizability is the cause of this effect.

1. Introduction

Atomic clusters are aggregates of a few up to many thousands of atoms that show properties intermediate between those of individual molecules and bulk matter. They are interesting both from the fundamental point of view - providing us with relatively simple model systems for studying the dynamical behavior when many degrees of freedom are involved - and from the more applied point of view. Since many properties, e.g., optical, electronic and magnetic, change dramatically with cluster size, clusters are becoming extremely interesting for a multitude of applications and provide a bridge between atomic, molecular and mesoscopic physics.

[1] **Varon Visiting Professor**

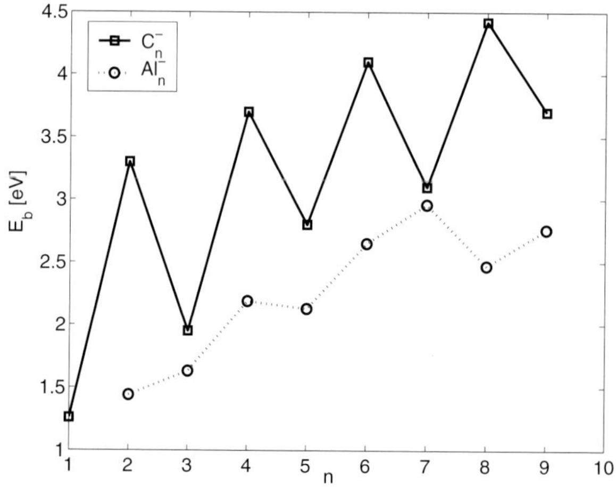

Figure 1. Calculated electron binding energy for negative carbon and aluminum clusters[19, 21].

Several properties of clusters have already been investigated systematically, such as the connection between the electronic structure and the mass spectra intensities, the size dependence of the melting point, and the optical response[1, 2, 3]. Experiments related to the size dependence of collision induced fragmentation and ionization have also been conducted[4, 5].

In the following, we present results for the size dependence of the electron impact detachment cross sections of negatively charged clusters. Electron impact detachment has been extensively studied over a large range of impact energies and for a large variety of atomic and molecular species [6, 7, 8, 9, 10, 11, 12, 13, 14]. Additionally, several theoretical works have been dedicated to this process, focusing both on the threshold energy region as well as on impact energies larger than the electron binding energy E_b[7, 11, 12]. These studies have demonstrated that the most important parameter determining both the threshold behavior and the magnitude of the cross section is E_b. Using both classical and quantum mechanical arguments, a simple scaling law has been derived, namely that the cross section is inversely proportional to the square of the binding energy of the least bound electron ($\sigma \propto 1/E_b^2$)[7, 12, 14].

Many of the most recent experiments related to electron impact detachment have been carried out using the heavy-ion storage ring technique. With this method, the ions can be stored for a time which is long enough for relaxation of the internal excitation (electronic and nuclear), which is usually produced in the ion source. The work presented here was carried out in a recently developed system combining a linear electrostatic ion beam trap and an electron target[15, 16, 17, 18], allowing studies of electron impact processes for fully relaxed ions of any mass. The experiments were performed with carbon (C_n^-, $1 \leq n \leq 9$) and aluminum (Al_n^-, $1 < n \leq 5$) clusters. These clusters were chosen because they differ dramatically in their structure: linear chain for the former [19, 20] and compact planar geometry for the latter [21]. The calculated electron binding energies for these species are shown in Fig. 1, and they exhibit the well known even-odd oscillation for C_n^- [19].

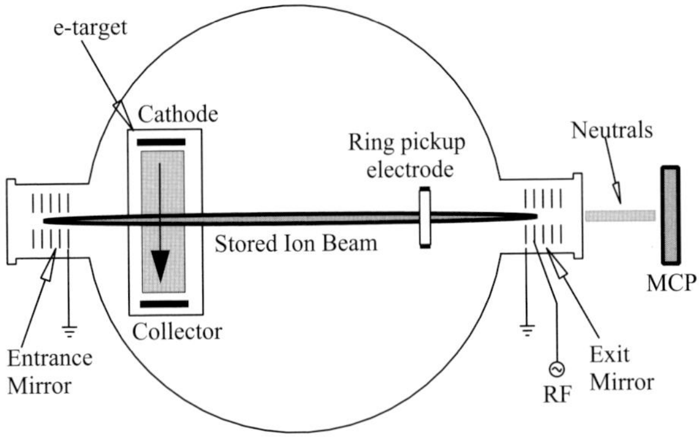

Figure 2. Schematic view of the experimental system. The diameter of the chamber is 451 mm.

2. Experimental setup

The trap used in this work is an enhanced version of the electrostatic trap described in Refs. [15, 16], and is described in detail elsewhere[18]. The trap has two electrostatic mirrors facing each other, each made of 8 electrodes, identical to those described in Refs. [15, 16]; the distance between the innermost electrodes is 490 mm (see Fig. 2). Ions are injected into the trap by lowering the voltages on the entrance electrodes (on the left hand side of Fig. 2). More details about the operation of the trap and its characteristics can be found in Refs. [22, 23]. The low pressure in the trap ($\sim 1\times10^{-11}$ Torr), is maintained by a cryopump and three non-evaporable getter pumps. A pickup electrode, located between the two mirrors (see Fig. 2), is used to measure the number of ions in the trap when operating in a bunch mode. The signal induced on the electrode by the passage of a bunch of ions is amplified by a charge sensitive amplifier, and fed to a digital scope. Fourier analysis of the signal gives both the frequency of oscillation and the amplitude which is proportional to the number of stored ions. Bunching is achieved by applying a small RF voltage to one of the electrodes of the exit mirror, with a frequency that is equal to the natural oscillation frequency of the particles in the trap (of the order of 100 kHz) and an amplitude of 10 V peak-to-peak. The amplitude of the RF signal is raised slowly ($\sim$120 ms) after the injection in order to reduce ion losses.

The electron target is located in the field-free region between the two electrostatic mirrors, closer to the entrance electrodes. It produces a flat, ribbon-like electron beam, $\sim$60 mm wide in the interaction region (see Fig. 2). Neutral products from the electron-ion interaction region leave the trap on straight trajectories and are detected by a multichannel plate (MCP) detector behind the trap. The electron beam energy is adjustable from $\sim$50 eV down to $\sim$7 eV, with current ranging from 500 to 20 μA, respectively. In order to avoid the presence of electric and magnetic fields, which would disturb the trajectories of the trapped ions, the electron beam is guided and focused only by electrostatic fields, and the interaction between electrons and ions takes place between two large grounded electrodes. More details about the electron target can be found in Ref. [18]

The measurement cycle starts by opening the chopper at the ion source for several tens of microseconds. (The DC beam from the sputter source is chopped to reduce the background of neutrals produced before the trap [16]). These ions, usually at an energy of 4.2 keV, are

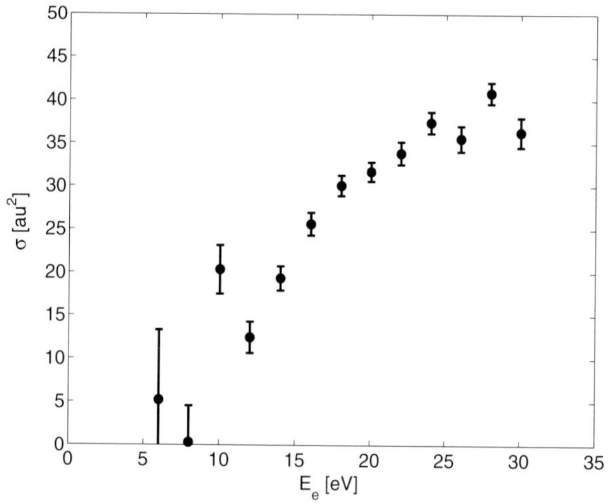

Figure 3. Electron impact detachment cross section of C_5^-. The data is normalized to the cross section of C_2^- at 20 eV[24].

mass selected by both a 20° magnet near the source and a 90° magnet about halfway between the source and the ion trap. Once the ion trap is filled (as can be verified with the pickup electrode), the voltages of the entrance mirror electrodes are raised and the beam is trapped. The ions then oscillate between the two mirrors, with a lifetime of the order of 500 ms at a residual pressure of $\sim 2\times 10^{-11}$ Torr. A cooling time of 100 ms following the injection was used to allow for electronic and vibrational relaxation of the clusters. This relaxation could be monitored through the non-exponential decay of the number of ions

Neutral particles produced either by collisions with the residual gas (background) or by the electron beam (signal) and moving downstream are counted using the MCP detector (see Fig. 2). In order to separate the signal from the background, the electron beam was modulated at a rate of 20 Hz, 20% of the cycle with electrons "on" and 80% with electrons "off". A measuring cycle comprises 10 injections at the same electron energy, plus one injection for normalization purposes (where the RF is switched on, and the number of trapped ions is measured up to a constant;see Ref. [18]). For each injection, the beam is trapped for a time that is comparable to the beam lifetime, and an additional 200 ms is added for background measurement (no beam) after grounding the entrance electrodes.

3. Relative cross section measurement

We have measured the relative cross section for electron impact detachment of C_n^- $(1 \le n \le 9)$ and Al_n^- $(1 < n \le 5)$ for electron energies between 7 to 30 eV. As the number of trapped ions is measured only up to a constant, we have normalized our results to the known value of the cross section for C_2^- at an electron energy of E_e=20 eV, as measured at the ASTRID storage ring[24].

As an example, the measured cross section for the electron impact detachment of C_5^- is shown in Fig. 3. The data shows that the cross section increases as a function of the electron

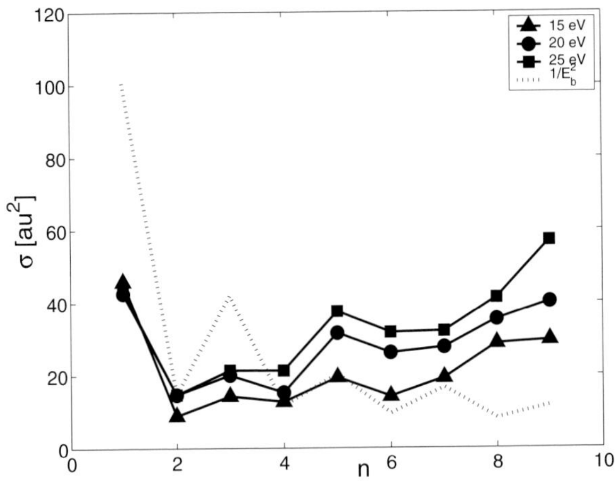

Figure 4. Measured electron impact detachment cross section as a function of carbon cluster size, for three different electron energies (full lines) and the scaling law predictions (dotted line) normalized to the experimental results at 20 eV for C_2^-.

energy. Similar spectra were obtained for all the species. Measurement below 6 eV were not possible due to the low electron beam current. Hence, possible resonances due to the presence of doubly negative clusters could not be observed, except for the case of C_2^-, where a clear resonance structure was observed[18], similar to the one measured at ASTRID[24].

In order to extract any possible size effect, the cross sections for the different clusters were compared at three different energies, away from the threshold region. The results are shown in Fig. 4 for carbon clusters and in Fig. 5 for aluminum clusters.

We first concentrate on the carbon cluster results, which are shown in Fig. 4 for three different electron energies. For the three sets of data, the trend is similar, namely an odd-even oscillation, which follows nicely the binding energy (see Fig. 1), i.e., clusters having a larger binding energy (the even-n) have a smaller cross section. This is in agreement with the $1/E_b^2$ scaling predicted theoretically based on either classical [7, 12] or quantum mechanical [11] arguments. However, an average increase of the cross section as a function of n (for $n \geq 2$) is observed, which is in contradiction with the scaling law, as the binding energy increases on average as well (see Fig. 1). Thus, the carbon cluster data shows one trend and its opposite at the same time.

The results for the aluminum clusters (Fig. 5), in contrast to the carbon clusters, are in good agreement with the scaling-law predictions in the range where measurements were possible, and the overall trend seems to follow the expected one, i.e., a decreasing cross section for higher electron binding energies. Measurements for clusters larger than n=5 were not feasible due to the low current from the ion source.

4. Discussion

The results presented in the previous Section show that the electron detachment cross section increases on average with cluster size for carbon, while it decreases for aluminum (at least

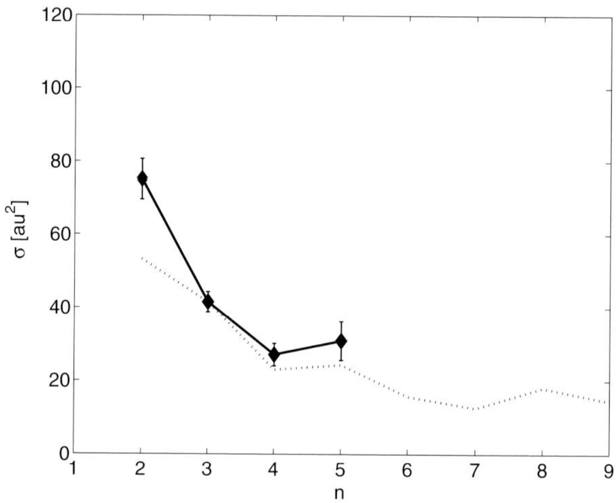

Figure 5. Measured electron impact detachment cross section as a function of aluminum cluster size (solid line) and scaling law predictions (dotted line).

for the range of clusters which could be measured). For both cases, the electron binding energy is increasing (see Fig. 1). Since the known scaling law is only based on the electron binding energy, this data demonstrates that at least one additional parameter is playing an important role in determining the size of the cross section. As already discussed by Andersen[14], Pindzola[10], and Ostrovsky and Taulberg[8], the next important parameter, after the binding energy, is the cluster polarizability. The importance of the polarizability can be understood directly from Robicheaux's argument [11] that electron detachment mainly occurs at the point of closest approach when the incoming electron has minimum kinetic energy. The attractive nature of the polarisation potential will reduce the distance of closest approach, thus resulting in larger distortion of the initial wave function and enhanced electron detachment. Several theoretical works have included a polarization potential in their calculations [8, 10], but the contribution of the polarizability to the electron detachment process has never been studied systematically, mostly due to the lack of relevant data.

It is well known that in the classical approximation, the polarizability is proportional to the volume occupied by the electrons [25, 26]. Since the structures of the carbon and aluminum clusters are very different, it can be expected that the polarizability will show a very different dependence on n. More explicitly, we expect the polarizability of the negative carbon clusters to increase with size more rapidly than for the aluminum species. Since there are no calculations for the polarizability of negative carbon and aluminum clusters, we have begun to perform such calculations, and in the following we present some preliminary results for the carbon clusters.

4.1. Polarizability of negative carbon clusters

Our first-principle calculations for the polarizabilities of carbon chain anions are based on density functional theory (DFT), where the local density approximation of Cepreley and Alder [27] is employed in conjunction with Troullier-Martins norm-conserving pseudopotentials [28]. All

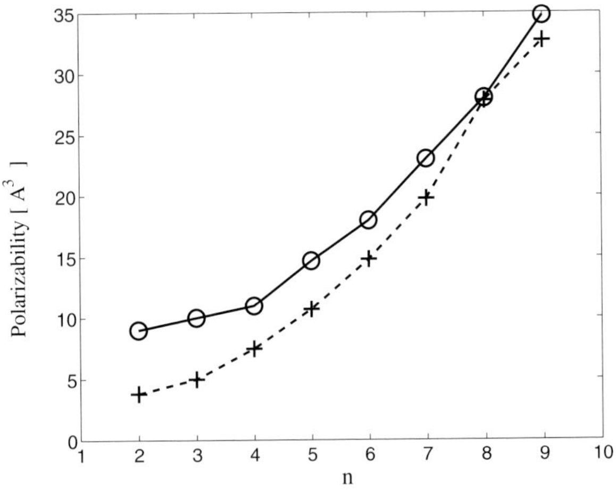

Figure 6. Calculated average polarizability for carbon cluster anions (circles - this work) and neutrals (plus signs - after ref. [20]).

calculations were performed using PARSEC - a software suite that solves the Kohn-Sham equations on a real-space grid by using the higher-order finite difference method [29].

After optimizing the chain geometry, polarizabilities were computed using the finite-field approach, which solves the Kohn-Sham equations with and without a small applied electric field [30]. The polarizability tensor is then calculated explicitly, via finite differences, as the first derivative of the dipole or the second derivative of the total energy. The average polarizability (which is measured experimentally) is then the average of the diagonal elements of this tensor. This approach (a detailed account of which has been given elsewhere [30]) has been shown previously to allow quantitative interpretation of experiments for both metallic [30] and semiconducting [31] clusters.

Preliminary results for the average polarizability of the carbon chain anions, C_n^- (n=2 to 9), are given in Fig. 6. For comparison, the same figure shows the many-body average polarizability calculations of Abdurahman *et al.* [20] for the carbon chain neutrals. We find that the anion polarizability is always larger than that of the corresponding neutral. This is because the extra electron is relatively weakly bound, and easy to polarize. This effect is very large for the dimer (where the anion/neutral polarizability ratio is $\sim$2.3), but decreases rapidly, such that for C_9 the ratio is only $\sim$1.06. Importantly, this difference cannot be attributed to the different computational techniques used for the neutrals and anions. This is because for the C_3 to C_7 neutrals the many-body calculations shown in Fig. 6 agree well with the DFT calculations of Fuentealba [32].

Most importantly, the results show that the anion polarizability increases monotonously with chain size, just like the neutral polarizability. This provides clear-cut computational support to the above-given interpretation of the detachment cross-section trends. Additional calculations for the polarization of the Al_n^- clusters are being carried out. The results of these calculations, together with the experimental results presented above, could serve as a basis for a more complete theory of electron impact detachment, as well as for a more detailled understanding of the influence of the polarization on electron induced processes.

Acknowledgments

This work is supported by the Israel Science Foundation and by the German Israel Foundation for Scientific Research (GIF) under contract I-707-55.7/2001. IBI is partly supported by the Chemical Sciences, Geosciences and Biosciences Division, Office of Basic Energy Sciences, Office of Science, U.S. Department of Energy.

References

[1] T. P. Martin, Phys. Rep. **273**, 199 (1996).

[2] M. Schmidt, *et al.*, Nature **393**, 238 (1998).

[3] T. Reiners, *et al.*, Phys. Rev. Lett. **74**, 1558 (1995).

[4] B. Farizon *et al.*, Eur. Phys. J. D **5**, 5 (1999).

[5] D. Hathiramani *et al.* Chem. Phys. Lett. **319**, 13 (2000).

[6] L. H. Andersen, T. Andersen,and P. Hvelplund, in *adv. in Atomic Molecular, and Optical Physics* edited by B. Bederson and H. Walther, **38**, 155 (Academic Press, New York 1998).

[7] V. A. Esaulov, J. Phys. B **13**, 1625 (1980); and references therein.

[8] V. N. Ostrovsky and K. Taulbjerg, J. Phys. B **29**, 2573 (1996).

[9] J. T. Lin, T. F. Jiang and C. D. Lin, J. Phys. B **29**, 6175 (1996).

[10] M. S. Pindzola, Phys. Rev. A **54**, 3671 (1996).

[11] F. Robicheaux, Phys. Rev. Lett. **82**, 707 (1999) and Phys. Rev. A **60**, 1206 (1999).

[12] L. Vejby-Christensen *et al.*, Phys. Rev. A **53**, 2371 (1996).

[13] J. M. Rost, Phys. Rev. Lett. **82**, 1652 (1999).

[14] L. H. Andersen, in *Photonic, Electronic, and Atomic Collisions - 22nd Int'l Conf.*, Edited by J. Burgdørfer, J.S. Cohen, S. Datz, C.R. Vane, (Rinton Press, Princeton 2002), p. 292.

[15] D. Zajfman *et al.*, Phys. Rev. A **55**, R1577 (1997).

[16] M. Dahan *et al.*, Rev. Sci. Instrum. **69**, 76 (1998).

[17] P. D. Witte, PhD Thesis, unpublished, University of Heidelberg 2002 (www.ub.uni-heidelberg.de/archiv/2977).

[18] O. Heber *et al.*, submitted for publication in Rev. Sci. Instrum. (2004).

[19] S. Yang *et al.*, Chem. Phys. Lett. **144**, 431 (1988).

[20] A. Abdurahman, A. Shukla, and G. Seifert, Phys. Rev. B **66**, 155423 (2002).

[21] B.K. Rao and P. Jena, J. Chem. Phys. **111**, 1890 (1999).

[22] H. B. Pedersen *et al.*, Phys. Rev. A **65**, 42703 (2002).

[23] H. B. Pedersen *et al.*, Phys. Rev. A **65**, 42704 (2002).

[24] H. B. Pedersen, *et al.*, Phys. Rev. A **60**, 2882 (1999).

[25] S. P. Apell *et al.*, Int. J. Quant. Chem. **86**, 35 (2002).

[26] M. Bianchetti *et al.*, Phys. Rep. **357**, 459 (2002).

[27] D. M. Ceperley and B. J. Alder, Phys. Rev. Lett. **45**, 566 (1980).

[28] N. Troullier and J. L. Martins, Phys. Rev. B **43**, 1993 (1991).

[29] J. R. Chelikowsky *et al.*, in C. Le Bris, Editor, *Handbook of Numerical Analysis - Volume X: Computational Chemistry* (Elsevier, Amsterdam, 2003).

[30] L. Kronik *et al.*, J. Chem. Phys. **115**, 4322 (2001).

[31] I. Vasiliev *et al.*, Phys. Rev. Lett. **78**, 4805 (1997).

[32] P. Fuentealba, Phys. Rev. A **58**, 4232 (1998).

Conference Programme

OPENING SESSION

ELECTRON IMPACT IONIZATION OF ATOMS AND MOLECULES I
(Chair : Azzedine Lahmam-Bennani)

Michaël Coplan (USA)
Identification of double ionization mechanisms, results bearing on electron correlation measurements

Andrew Murray (United Kingdom)
(e,2e) measurements in calcium at low energy

Ugo Ancarani (France)
Role of the Ground state in (e,3e) processes on helium

ELECTRON IMPACT IONIZATION OF ATOMS AND MOLECULES II
(Chair : Uwe Becker)

Nikolay Kabachnik (Russia)
Modern studies of anisotropy and polarization in Auger-electron emission

Stéphane Rioual (France)
Coherence effects in Auger photoelectron coincidence experiments

POSTER SESSION

ELECTRON IMPACT IONIZATION OF ATOMS AND MOLECULES III
(Chair : Claude Dal Cappello)

Fabrice Catoire (France)
Correlations in Auger process under electron impact ionization of Argon

Alberto De Fanis (Japan)
Doppler-free resonant Auger Raman spectroscopy on atoms and molecules at SP ring -8

Carlos Stia (Argentina)
Interference effects in ionization of H_2 molecules by fast electron impact

Maarten Vos (Australia)
Elastic scattering at high momentum transfer, a possible probe of quantum entanglement

HEAVY PARTICLE AND PHOTON IMPACT IONIZATION OF ATOMS AND MOLECULES I
(Chair : Laurence Malegat)

Joachim Ullrich (Germany)
Multiple ionization of atoms in strong fields

Alexei Grum-Grzhimaïlo (Russia)
Complete experiments in photoionization with excitation : theory and practice

HEAVY PARTICLE AND PHOTON IMPACT IONIZATION OF ATOMS AND MOLECULES II
(Chair : Peter Lambropoulos)

John Eland (United Kingdom)
Double photoionization of molecules via T.O.F. electron-electron coincidence experiments

Mohand Brouri (Belgium)
Simultaneous versus sequential two photon double electron ejection from He^-

Marc Vrakking (The Netherlands)
Strongly driven electrons

POSTER SESSION

HEAVY PARTICLE AND PHOTON IMPACT IONIZATION OF ATOMS AND MOLECULES III
(Chair : Henri Bachau)

Ulrich Kleiman (U.S.A.)
Multiple photoionization of atoms and ions using the time-dependent close-coupling method

Till Jahnke (Germany)
Interatomic Coulombic decay in fixed-in-space Ne-dimers

HEAVY PARTICLE AND PHOTON IMPACT IONIZATION OF ATOMS AND MOLECULES IV
(Chair : Hugo van der Hart)

Patricia Selles (France)
Ab-initio calculations of the double photoionization cross sections for helium down to 0.1 ev from threshold

Robert Moshammer (Germany)
Single and double ionization by ion impact

Robin Shakeshaft (U.S.A.)
An integral representation of the Coulomb Green function applied to two-electron escape from H^- *by photoabsorption*

ELECTRON AND PHOTON IMPACT ON CLUSTERS
(Chair : Jan Posthumus)

Hubertus Wabnitz (France)
Non-linear processes in clusters with FEL radiation

Ulf Saalmann (Germany)
Interaction of strong X-ray lasers with rare-gas clusters

Daniel Zajfman (Israel)
Electron impact detachment of small negative clusters

GENERAL DISCUSSION AND CLOSURE

Poster Session

Polarization effects in ionization-excitation of helium
K Bartschat and N Andersen

On the helium Schrödinger equation singularities and non-separability
T Montagnese and L U Ancarani

A step towards a unified treatment of doubly excited and double continuum states
C Bouri and L Malegat

Single and double photoionization of H^-
E Foumouo ,G Edah, and B Piraux

Fully differential cross sections of Helium double ionization in intense laser pulses
A Staudte, D Zeidler, M Weckenbrock, Th Weber, M Schöffler, M Meckel, S Kammer, M Smolarski, O Jagutzki, V R Bhardwaj, D M Rayner, D M Villeneuve, P B Corkum and R Dörner

Test of the Pluvinage wave function for the helium ground state
S Jones, J H Macek and D H Madison

Analysis of the helium double photoionization using an improved parametrization with variable phase-shift and shake-off contribution
S Cvejanovic, J Viefhaus, G Prümper, M Wiedenhöft, N Berrah and U Becker

Dynamics of core-excited molecules by synchrotron radiation-induced electron and ion spectroscopies
M N Piancastelli

Multiphoton processes in helium investigated using the R-matrix Floquet approach
H van der Hart and P Bingham

Multiphoton single and double ionization of two-active electron systems with a time-dependent spectral method
E Foumouo, G Lagmago Kamta and B Piraux

Ab-initio calculation of C1s photoionization spectra of C_2H_2
D Embriaco, R Colle, M Massini, S Simonucci, S Taioli

Ab-initio calculation of angle-resolved resonant Auger spectra of C_2H_2
D Embriaco, R Colle, M Massini, S Simonucci and S Taioli

DWIA and 3DW calculations for electron-impact ionization of molecules
J Gao, DH Madison and JL Peacher

Intense field - molecule interactions: comparing H_2^+ and H_2 targets
J H Posthumus, N de Ruette, B Fabre and X Urbain

Evidence of new resonant states in the H^2 molecule
J Fernández, F Martín, T Aoto, Y Hikosaka, R I Hall and K Ito

Dissociative photoionization of HeH^+
J Fernández and F Martín

Two-photon dissociative ionization of H_2^+ in perturbative and non-perturbative regimes
A Palacios, S Barmaki, H Bachau and F Martín

Atomic and molecular double photoionization calculations using exterior complex scaling
D A Horner, W I Vanroose, F Martín, Th N Rescigno and C W McCurdy

Ionization dynamics in ultra-short laser pulses: coincident recoil ion - electron spectroscopy
CD Schröter, A Rudenko, B Feuerstein, K Zrost, V L B de Jesus, T Ergler, R Moshammer and J Ullrich

Attosecond pulses to explore the electron-electron correlation in atoms
E Foumouo, B Piraux, S Laulan and H Bachau

Resonant ionization of D_2 by high energy electron impact
J Fernández, F Martín, O Kamalou, P Sobocinski, J-Y Chesnel, D Martina and F Fremont

Double electron emission processes in atoms, molecules and fullerenes
J Viefhaus

A reaction microscope for low energy electron collision experiments
A Dorn, M Dürr, J Steinmann, C Höhr, C-D Schröter, R Moshammer and J Ullrich

Theory of (e,2e) reactions
A T Stelbovics, A S Kadyrov, A M Mukhamedzhanov and I Bray

New numerical method for single and double ionization of two-electron systems by fast electrons
V V Serov, V L Derbov, B Joulakian and S I Vinitsky

Double ionization of helium by electron impact
C Dal Cappello and S Elazzouzi

QED theory of multiphoton resonances in spectra of atoms in a strong laser field
A V Glushkov

QED calculation of characteristics of the elementary processes in collisionally pumped plasma and searching the plasma parameters for X-ray lasing
A V Glushkov

Stochastic kinetics modeling of the multi-photon photo excitement and dissociation for CF_3BR molecules with account for energy transfer, rotational and V-T relaxation
A V Glushkov

New optimal schemes of the laser photoionization method of preparing the films of pure composition at atomic level
S V Malinovskaya

QED calculation of the nuclear gamma transition spectra in atomic and molecular systems: new laser-electron nuclear effects
S V Malinovskaya

(e,2e) Measurements on xenon: the fine-structure effect revisited
J Lower, R Panajotovic, E Weigold, D H Madison and J Parks

Electron-impact ionization from n=2 levels of atomic hydrogen
Ph L Bartlett and A T Stelbovics

Distorting and higher-order effects for electron impact ionization of helium
Zhangjin Chen, D H Madison, C T Whelan and H R J Walters

(e,3-1e) reactions at large momentum transfer: analysis of recent experimental results
P S Vinitsky, O Chuluunbaatar, Y Khajuria, K A Kouzakov, Yu V Popov, M Takahashi, Y Udagawa and N Watanabe

(e,3-1e) reactions at large momentum transfer: details of the experiment and comparison with theory
Y Khajuria, N Watanabe, M Takahashi, Y Udagawa, P S Vinitski, O Chulluunbaatar, K A Kouzakov and Yu V Popov

Dynamics of Landau and Zeeman wave packets in a time- dependent electric field
O Chuluunbaatar, A A Gusev, M S Kaschiev, K A Kousakov, Yu V Popov, and S I Vinitsky

Benchmark three-body models
O Chuluunbaatar, A A Gusev, M S Kaschiev, S Y Larsen and S I Vinitsky

Observation of molecular frame (e,2e) cross section using an electron-electron -fragment ion triple coincidence apparatus
M Takahashi, N Watanabe, Y Khajuria, Y Udagawa and J H D Eland

Importance of relativistic effects in spin-asymmetry measurements for electron-xenon scattering
A Prideaux and D H Madison

Electron impact ionization of the hydrogen molecule at intermediate energies
D Milne-Brownlie and Birgit Lohmann

Double ionisation of H_2 molecule by electron impact: contribution of the second Born term
A Mansouri, C Dal Cappello, S Houamer, I Charpentier and A Lahmam-Bennani

Analysis of interference patterns in photoionization of H_2 at high incident energy
O A Fojon, J Fernández, A Palacios, R D Rivarola and F Martín

Spin-polarized (e,2e) spectroscopy of ferromagnetic/metal and ferromagnetic/semi conductor systems
S N Samarin, O M Artamonov, A D Sergeant, A A Suvorova and J F Williams

Scattering of spin-polarized electrons from W(110) by (e,2e) spectroscopy
S N Samarin, O M Artamonov, A D Sergeant and J F Williams

Valence satellite of neon by electron momentum spectroscopy
N Watanabe, Y Khajuria, M Takahashi and Y Udagawa

EMS and DFT investigation of the Cu/Si interface
K L Nixon, C Bowles, M J Ford and M Vos

Core hole alignment and induced dichroïsm in angle resolved Auger-photoelectron coïncidence spectroscopy
G Stefani, R Gotter, A Rocco, F Offi, F Da Pieve, S Iacobucci, A Liscio, H Yao and R A Bartynski

Green's function studies of the valence ionization spectra of Norbornane, norbornene, norbornadiene, stella-2,6-dione and stella-2,6-diene compounds
S Knippenberg, J-P Francois and M S Deleuze

Laser assisted (e,2e) collision
A Dorn, C Höhr, M Dürr, J Steinmann, G D Schröter, R Moshammer and J Ullrich

(e,3e) on helium in collisions at relativistic energies
B Najjari, A B Voitkiv, A Dorn and J Ullrich

Four-body treatment of single ionization of helium by protons
M Foster, D H Madison, J L Peacher and M Schulz

Fully differential cross section for impact ionization of helium by 2 Mev/amu C^{6+} ions
M Foster, D H Madison, J L Peacher, M Schulz, S Jones, D Fischer, R Moshammer and J Ullrich

Multiple photoionization of clusters and fullerenes
O Kidun, N Fominykh and J Berakdar

List of Participants

ANCARANI, Lorenzo Ugo
LPMC, Université de Metz
1 Bd Arago
F-57078 Metz Cedex 3,
France
ancarani@sciences.univ-metz.fr

BACHAU, Henri
CELIA, Université de Bordeaux I
351 Cours de la libération
F-33405 Talence,
France
bachau @ celia.u-bordeaux1.fr

BARTSCHAT, Klaus
Drake University
Physics Department
2507 University Avenue
Des Moines, IA 50311
USA
klaus.bartschat@drake.edu

BECKER, Uwe
Fritz-Haber-Institut
Faradayweg 4-6
D-14195 Berlin
Deutschland
becker_u@fhi-berlin.mpg.de

BOURI, Celsus
Lab. LIXAM
Université de Paris Sud,
UMR 8624 du CNRS, Bât. 350
F-91405 Orsay
France
bouricelsus@yahoo.fr
and CEPAMOQ
University of Douala
Faculté des Sciences
P.O. Box 8580
Douala
Cameroon

BROURI, Mohand
Université catholique de Louvain, FYAM
chemin du cyclotron 2
B-1348 Louvain-la-Neuve
Belgium
brouri@fyam.ucl.ac.be

CATOIRE, Fabrice
Université de Paris-Sud, bât. 351
F-91405 Orsay Cedex
France
catoire@lcam.u-psud.fr

COPLAN, Michael
University of Maryland
IPST CSS Building
College Park, MD 20742
USA
coplan@glue.umd.edu

CROWE, Albert
University of Newcastle
School of Natural Sc.(Physics)
Herschel Building
Newcastle upon Tyne NE1 7RU
United Kingdom
albert.crowe@ncl.ac.uk

CVEJANOVIC, Slobodan
University of Rijeka
Medical faculty
Brace Branchetta 20
51000 Rijeka
Croatia
cvejanovic@medri.hr

DA PIEVE, Fabiana
University Roma Tre
Via della Vasca Navale 84
I-00146 Roma
Italy
dapieve@fis.uniroma3.it

DAL CAPPELLO, Claude
Université de Metz
LPMC, Institut de physique
1 rue Arago
F-57078 Metz Cedex 3
France
cappello@sciences.univ-metz.fr

DE FANIS, Alberto
JASRI / Spring-8
1-1-1 Kouto Mikazuki, Sayo-gun,
Hyogo 679-5198
Japan
defanis@spring8.or.jp

DEFRANCE, Pierre
Université catholique de Louvain, FYAM
chemin du cyclotron 2
B-1348 Louvain-la-Neuve
Belgium
defrance@fyam.ucl.ac.be

DELEUZE, Michaël
Limburgs Universitair Centrum
B-3590 Diepenbeek
Belgium
michael.deleuze@luc.ac.be

DORN, Alexander
Max Planck Institut for Nuclear Physics
Saupfercheckweg 1
D-69117 Heidelberg
Deutschland
dornalex@mpi-hd.mpg.de

ELAND, John
Oxford University
Physical Chemistry Lab.
South parks road
Oxford OX1 3QZ
United Kingdom
eland@physchem.ox.ac.uk

ELAZZOUZI, Saïd
Université de Metz
LPMC, Institut de physique
1 rue Arago
F-57078 Metz Cedex 3
France
elazzouzi@sciences.univ-metz.fr

EMBRIACO, Davide
Scuola Normale Superiore
Piazza dei Cavalieri 7
I-56100 Pisa
Italy
embriaco@sns.it

FABRE, Baptiste
Université catholique de Louvain, FYAM
chemin du cyclotron 2
B1348 Louvain-la-Neuve
Belgium
fabre@fyam.ucl.ac.be

FERNANDEZ, Jorge
Universidad Autonoma de Madrid
Ctra. De Colmenar km 15
E28049 Madrid
Spain
j.fernandez@uam.es
FOUMOUO, Emmanuel
Université catholique de Louvain, FYAM
chemin du cyclotron 2
B-1348 Louvain-la-Neuve
Belgium
foumouo@fyam.ucl.ac.be

FRANCOIS, Jean-Pierre
Limburgs Universitair Centrum
B-3590 Diepenbeek
Belgium

GLUSHKOV, Alexander
Atom.-Nucl.-Laser Spectroscopy Centre &
Inst.of Appl. Math.
OEU Odessa University
PO Box 116
Odessa-9
65009 Ukraine
glushkov@paco.net

GRUM-GRZHIMAILO, Alexei
Institute of Nuclear Physics
Moscow State University
Moscow 119992
Russia
algrgr1492@yahoo.com
grum@anna19.npi.msu.su

HORNER, Daniel
Lawrence Berkeley National Laboratories
1 cyclotron road, Mailstop 50F
Berkeley, CA 94720,
USA
dahorner@lbl.gov

JAHNKE, Till
Institut für Kernphysik
University of Frankfurt,
August Euler str. 4-6
D-60486 Frankfurt
Deutschland
jahnke@hsb.uni-frankfurt.de

KABACHNIK, Nikolay
Moscow State University
Moscow 119992
Russia
and Fakultät für Physik
Universität Bielefeld
D-33615 Bielefeld
Deutschland
nkabach@mail.ru

KHAJURIA, Yugal
Institute for Molecular Science 38
Nishigo-Naka
Myodaiji - Okazaki 444-8585
Japan
khajuria@ims.ac.jp

KIDUN, Oleg
Max-Planck Institute of Microstructure Physics
Weinberg 2
Halle
Deutschland
okidun@mpi-halle.de

KLEIMAN, Ulrich
Auburn University
Department of Physics
206 Allison Laboratory
Auburn AL36849-5311
USA
kleiman@physics.auburn.edu

KNIPPENBERG, Stefan
Limburgs Universitair Centrum
B-3590 Diepenbeek
Belgium
stefan.knippenberg@luc.ac.be

LAHMAM-BENNANI, Azzedine
Université de Paris-Sud XI
LCAM, Bât. 351
F91405 Orsay Cedex
France
azzedine.l-bennani@lcam.u-psud.fr

LAMBROPOULOS, Panayotis
University of Crete, Physics
Heraklion
Crete 71003
Greece
labro@iesl.forth.gr

LECOINTRE, Julien
Université catholique de Louvain, FYAM
chemin du cyclotron 2
B1348 Louvain-la-Neuve
Belgium
lecointre@fyam.ucl.ac.be

LOHMANN, Birgit
Griffith University
Nathan
Queensland
Australia
b.lohmann@griffith.edu.au

LOWER, Julian
Australian National University
Canberra ACT 0200
Australia
Julian.Lower@anu.edu.au

MADISON, Don
University of Missouri-Rolla
Laboratory for Atomic and Molecular Research
1870 Miner Circle - 107 Physics
Rolla MO 65409 0640
USA
madison@umr.edu

MALEGAT, Laurence
Université Paris Sud 11
LIXAM, bâtiment 350
F-91405 Orsay
France
Laurence.Malegat@lixam.u-psud.fr

MALINOVSKAYA, Svetlana
Odessa University
P.O. Box 116
Odessa-9
65009 Ukraine
trust@paco.net

MANSOURI, Abdelaziz
Université Ferhat Abbas
Faculté des Sciences
Département de physique
Setif
Algérie
azmansouri@yahoo.com

MOSHAMMER, Robert
Max Planck Institute for Nuclear Physics
Saupfercheckweg 1
D69117 Heidelberg
Deutschland
r. moshammer@mpi-hd.mpg.de

MURRAY, Andrew
Dpt Physics and Astronomy
University of Manchester
Manchester M13 9PL
United Kingdom
andrew.murray@man.ac.uk

NAJJARI, Bennaceur
Max Planck Institute
Saupfercheckweg 1,
D-69117 Heidelberg
Deutschland
najjari@mpi-hd.mpg.de

NIXON, Kate
Flinders University
School of chemistry,Physics & Earth sciences
GPO Box 2100
Adelaide
South Australia 5001
kate.nixon@flinders.edu.au

PALACIOS, Alicia
Departamento de Quimica C-9
Universidad Autonoma de Madrid
Carretera de Colmenar km 15
E-28049 Madrid
Spain
alicia.palacios@uam.es

PIANCASTELLI, Maria Novella
Physics Dept
Uppsala University
Box 530
S-75121 Uppsala
Sweden
maria-novella.piancastelli@fysik.uu.se

PIRAUX, Bernard
Université catholique de Louvain, FYAM
chemin du cyclotron 2
B-1348 Louvain-la-Neuve
Belgium
piraux@fyam.ucl.ac.be

POPOV, Yuri
Nuclear Physics Institute
Moscow State University
Moscow 119992
Russia
popov@srd.sinp.msu.ru

POSTHUMUS, Jan
Université catholique de Louvain, FYAM
chemin du cyclotron 2
B1348 Louvain-la-Neuve
Belgium
posthumus@fyam.ucl.ac.be

RIOUAL, Stéphane
Université de Brest
UFR Sciences, LCEA
6 av. V. le Gorgeu
F-29285 Brest
France
rioual@univ-brest.fr

RIVAROLA, Roberto
Instituto de Fisica Rosario
(CONICET-Universidad Nacional de Rosario)
Av. Pellegrini 250
2000 Rosario
Argentina
rivarola@fceia.unr.edu.ar or
rivarola@ifir.edu.ar

SAALMANN, Ulf
Max-Planck-Institute for Physics of
Complex Systems
Nöthnitzer st. 38
D-01187 Dresden
Deutschland
us@mpipks-dresden.mpg.de

SAMARIN, Sergey
School of physics
Univ. of Western Australia
35 Stirling Highway
Crawley WA 6009
Australia
samar@physics.uwa.edu.au

SCHRÖTER, Claus-Dieter
Max-Planck-Institut für Kernphysik
Saupfercheckweg 1
D69117 Heidelberg
Deutschland
c.schroeter@mpi-hd.mpg.de

SCHULZ, Michaël
University of Missouri-Rolla
Laboratory for Molecular and Optical Research
Physics Department
116 Physics
Rolla, Missouri 65409-0640
USA
schulz@umr.edu

SELLES, Patricia
Université Paris Sud, LIXAM
Bâtiment 350
F-91405 Orsay
France
patricia.selles@lixam.u-psud.fr

SEROV, Vladislav
Dept of theoretical and nuclear physics
Saratov State University
83 Astrakhanskaya street
410012 Saratov
Russia
vladislav_serov@mail.ru

SHAKESHAFT, Robin
University of Southern California
Physics Dept.
Los Angeles CA 90089-0484
USA
robins@usc.edu

STAICU-Casagrande, Elena M.
Université Paris Sud
LCAM, Bât. 351
F-91405 Orsay, France
casa_staicu@yahoo.fr

STAUDTE, André
University of Frankfurt
August-Euler-Str. 24
D-60486 Frankfurt
Deutschland
a.staudte@em.uni-frankfurt.de

STELBOVICS, Andris
Murdoch University
South street
Murdoch
Western Australia 6150
a.stelbovics@murdoch.edu.au

STIA, Carlos
Instituto de Fisica Rosario
av. Pellegrini 250
2000 Rosario
Argentina
stia@fceia.unr.edu.ar

SUVOROVA, Alexandra
Centre for microscopy and microanalysis
University of Western Australia
35 Stirling Highway
Crawley WA 6009
Australia
suvorova@cmm.uwa.edu.au

TAKAHASHI, Masahiko
Institute for Molecular Science
Okazaki 444-8585
Japan
masahiko@ims.ac.jp

TETCHOU, H.M.
Université catholique de Louvain, FYAM
chemin du cyclotron 2
B-1348 Louvain-la-Neuve
Belgium
tetchou@fyam.ucl.ac.be

ULLRICH, Joachim
Max-Planck-Institut für Kernphysik
Saupfercheckweg 1
D-69117 Heidelberg
Deutschland
joachim.ullrich@mpi-hd.mpg.de

URBAIN, Xavier
Université catholique de Louvain, FYAM
chemin du cyclotron 2
B-1348 Louvain-la-Neuve
Belgium
urbain@fyam.ucl.ac.be

van der HART, Hugo
Queen's University Belfast
David Bates Building
Belfast BT7 1NN
United Kingdom
h.vanderhart@qub.ac.uk

VOS, Maarten
Atomic and molecular physics laboratories
Research school of physical sciences and engineering
Australian National University
Canberra ACT 0200
Australia
maarten.vos@anu.edu.au

VRAKKING, Marc
Amolf FOM
Institute for Atomic and molecular physics
Kruislaan 407
NL-1098 SJ Amsterdam
The Netherlands
m.vrakking@amolf.nl

WABNITZ, Hubertus
Commissariat à l'Energie Atomique
DRECAM/SPAM
Centre d'Etudes de Saclay
F-91191 Gif-sur-Yvette
France
wabnitz@drecam.cea.fr

WATANABE, Noboru
Institute for Molecular Science
Myodaiji
Okazaki 444-8585
Japan
noboru@ims.ac.jp

WILLIAMS, Jim
The University of Western Australia
Physics Department
35 Stirling Highway
Crawley Perth. WA 6009
Australia
jfw@physics.uwa.edu.au

ZAJFMAN, Daniel
Weizmann Institute of Science
Rehovot 76100
Israel
daniel.zajfman@weizmann.ac.il

Author Index